Oxford Applied Mathematics and Computing Science Series

General Editors
J. Crank, H. G. Martin, D. M. Melluish

IAN ANDERSON
University of Glasgow

A first course in combinatorial mathematics

CLARENDON PRESS · OXFORD
1974

Oxford University Press, Ely House, London W. 1

GLASGOW   NEW YORK   TORONTO   MELBOURNE   WELLINGTON
CAPE TOWN   IBADAN   NAIROBI   DAR ES SALAAM   LUSAKA   ADDIS ABABA
DELHI   BOMBAY   CALCUTTA   MADRAS   KARACHI   LAHORE   DACCA
KUALA LUMPUR   SINGAPORE   HONG KONG   TOKYO

CASEBOUND ISBN 0 19 859616 2
PAPERBACK ISBN 0 19 859617 0

© OXFORD UNIVERSITY PRESS 1974

PRINTED IN GREAT BRITAIN
BY WILLIAM CLOWES & SONS LIMITED
LONDON, COLCHESTER AND BECCLES

Preface

IN 1857, the Rev. Thomas Kirkman presented to the Historic Society of Lancashire and Cheshire a paper which had nothing to do with either history or theology. The paper was concerned with a problem in what is now known as *combinatorial mathematics*, and Kirkman began by explaining the attraction of such problems. 'The elements to be combined in these questions', he remarked, 'have no property except that of diversity. They have no arithmetic value or capacity, except that they can be counted. No operation of addition, subtraction, multiplication, or division can be performed upon them. They can merely be combined.'

Combinatorial mathematics, then, is first of all concerned with counting the number of ways of arranging given objects in a prescribed way. The aim of the first part of this book is to introduce the reader to a working knowledge of the basic ideas and techniques of the subject. Chapters 2, 3, 4, and 5 contain essential technical know-how, Chapter 3 also being concerned with various aspects of *assignment* problems, beginning with the famous result of Philip Hall, and leading on to various applications. In an attempt to give a balanced view of the subject, the remaining chapters deal with configurations rather than techniques. The emphasis here is not so much on the question 'how many . . .?' but on the structure and properties of systems satisfying certain prescribed conditions. A study of block designs suggests applications to error-correcting codes, and in the final chapter a study of the Steiner system $S(5, 8, 24)$ leads on to the construction of the well-known Leech lattice in 24 dimensions. Thus some idea is given of how the properties of seemingly useless systems can be put to use in interesting applications.

This little book is not intended as an encyclopaedia for research workers: such people are well served already. It is intended as a textbook for anyone to work from who wishes to become acquainted with the flavour of the subject and the basic tools of the trade. The subject has come a long way since Kirkman's time, but it still remains an easily accessible area of mathematics, one which is becoming more and more widely used in other disciplines. The days are past when the calculus was thought to be the queen of applicable mathematics. But, despite its applications, the subject of this book is genuine mathematics in all its purity, and as such is worthy of study just for its own attractiveness and charm.

The reader who is new to the subject should not be put off by the lengthy list of references to mathematical papers at the end of the book. An understanding of the text in no way requires a study of these, but it is hoped that the reader will, as well as becoming aware of how contemporary the subject is, follow up one or two of these references and see for himself the type of work which is going on today.

A number of people have helped in the preparation of this book. My thanks go to Dr H. G. Martin and Dr J. R. Gillett who encouraged me to start writing, and to the Clarendon Press for their encouragement in the later stages. Finally I extend my gratitude to all those who first interested me in combinatorial ideas, and to all the mathematicians whose work in the subject is now the standard repertoire. To them be the glory.

Glasgow, January 1973   I.A.

Contents

viii   *Contents*

1 Introduction to basic ideas

TO embark upon a mass of unmotivated theory is perhaps the simplest and quickest way of losing the reader. This study of combinatorial ideas will therefore begin with a specific problem, and discover several methods for its solution, introducing the reader incidentally to some of the basic ideas on which the remainder of the book will build. Each of these approaches will be examined later in greater depth to see how they can be applied in other situations or modified to yield new techniques.

The specific problem to be considered is the following.

Problem

Suppose that each of k indistinguishable golf balls has to be coloured with any one of n given colours. How many different colourings are possible?

If x_1 denotes the number of balls coloured with the first colour, x_2 the number coloured with the second colour, and so on, the required number is the number of solutions of the equation

$$x_1 + x_2 + \cdots + x_n = k,$$

in non-negative integers $x_1, \ldots, x_n$. As this number will depend upon both n and k, denote it by $f(n, k)$.

Special cases

(a) If there is only one colour available (i.e. $n = 1$) the k balls can be coloured in only one way. Thus

$$f(1, k) = 1 \quad \text{for all } k \geqslant 1. \tag{1.1}$$

(b) If there are n colours but only one ball, there are n possible colourings. Thus

$$f(n, 1) = n \quad \text{for all } n \geqslant 1. \tag{1.2}$$

(c) If n and k are both very small, then $f(n, k)$ can be found without much difficulty. For example, suppose $n = k = 2$. If the colours are black and white, the balls can be coloured both black, both white, or one of each. Thus $f(2, 2) = 3$.

The problem is to find a general formula for $f(n, k)$. Three possible approaches will be mentioned.

First approach

It has already been pointed out that if n and k are both small then $f(n, k)$ can be found quite easily. This suggests that one method of attack would be to try to express $f(n, k)$ in terms of similar expressions with smaller values of n and k. For example, if it were known that

$$f(3, 2) = f(2, 2) + f(3, 1)$$

it could be deduced that

$$f(3, 2) = 3 + 3 = 6,$$

using results already known. So consider $f(n, k)$ and concentrate on the nth colour. When the k balls are coloured, this nth colour may or may not be used. If it is not used, there are in effect only $(n - 1)$ colours and so the colourings can be performed in $f(n - 1, k)$ ways. But if the nth colour is used, one ball can be coloured by it and then removed to leave $(k - 1)$ balls which can be coloured in $f(n, k - 1)$ ways. Thus

$$f(n, k) = f(n - 1, k) + f(n, k - 1). \qquad (1.3)$$

A relation such as (1.3) is called a *recurrence relation* and our ability to solve the original colouring problem now depends on our ability to solve recurrence relations. By solving (1.3) is meant finding the unique function f which satisfies (1.3) and the *boundary conditions* (1.1) and (1.2). In general, a recurrence relation has more than one possible solution, but the boundary conditions specify which of these is the required solution.

Recurrence relations are important in combinatorial mathematics and are widespread in their appearances. Perhaps the most famous of all occurs in the definition of the *Fibonacci sequence* $\{a_n\}$,

$$1, 2, 3, 5, 8, 13, 21, 34, 55, \ldots$$

given by the boundary conditions

$$a_1 = 1, \quad a_2 = 2,$$

and the recurrence relation

$$a_n = a_{n-1} + a_{n-2} \quad (n \geqslant 3).$$

From the third term onwards each is the sum of the two preceding terms. Solving this recurrence relation means obtaining a formula for the nth term a_n as a function of n. This will be done later.

Example 1.1. Solve the recurrence relation $a_n = na_{n-1}$, subject to the condition $a_1 = 1$.

Solution. The first few a_n can easily be found by working up from a_1. Thus $a_2 = 2a_1 = 2$, $a_3 = 3a_2 = 6$ and so on. But to find a formula for a_n it is best to start at the top and work down. Thus $a_n = na_{n-1}$ and similarly $a_{n-1} = (n-1)a_{n-2}$, so that

$$a_n = n(n-1)a_{n-2} = n(n-1)(n-2)a_{n-3}$$

and finally

$$a_n = n(n-1)(n-2) \ldots 2a_1$$
$$= n(n-1)(n-2) \ldots 2 . 1$$
$$= n!,$$

where $n!$ ('n factorial') is the product of the first n positive integers. Thus, for example,

$$a_3 = 3! = 3 . 2 . 1 = 6, \quad a_5 = 5! = 120.$$

Exercises 1.1

1. Show that $f(4, 2) = 10$ and $f(5, 3) = 35$.

2. Solve the recurrence relation $a_n = n^2 a_{n-1}$ given that $a_1 = 1$.

3. If $a_n = \dfrac{n-1}{n} a_{n-1}$ and $a_1 = 2$, find a_n.

4. If $a_n = a_{n-1} + n$, find a_n if (a) $a_1 = 1$, (b) $a_1 = 0$.

5. Suppose a row of n cages is given, and it is required to place k indistinguishable lions in them so that no cage contains more than one lion, and no two consecutive cages both contain a lion. Let $g(n, k)$ denote the number of ways in which this can be done. Prove that

    (a) $g(2k - 1, k) = 1$,
    (b) $g(n, k) = 0$ if $n < 2k - 1$,
    (c) $g(n, 1) = n$,
    (d) $g(n, k) = g(n - 2, k - 1) + g(n - 1, k)$ if $k \geqslant 2$.

    (Hint: consider whether or not the last cage contains a lion.) Deduce that

    (e) $g(6, 3) = 4$,
    (f) $g(2k, k) = g(2k - 2, k - 1) + 1$,
    (g) $g(2k, k) = k + 1$.

6. Suppose it is known that $t(n, n - 1) = 1$ and that

$$(n - k - 1)t(n, k) = k(n - 1)t(n, k + 1)$$

for each $k < n - 1$. Deduce that

$$t(n, k) = \frac{(n - 1)^{n-k-1}(n - 2)!}{(k - 1)!(n - k - 1)!}.$$

Second approach

Consider the effect of multiplying the following expressions together,

$$(1 + x + x^2 + x^3 + \cdots)(1 + x + x^2 + \cdots) \ldots (1 + x + x^2 + \cdots),$$

$$(1.4)$$

where there are n brackets. Different powers of x are obtained, each power occurring a number of times. For example, x^2 can be obtained in n ways by multiplying an x^2 term from one bracket by the terms 1 in each of the remaining brackets, and in yet more ways by multiplying two x terms from two of the brackets by the terms 1 in each of the remaining brackets.

More generally, how often will the term x^k appear, i.e. what will the coefficient of x^k be in the resulting expression? If the convention $x^0 = 1$ is followed, x^k can be obtained by choosing a term x^{t_1} in the first bracket, x^{t_2} in the second, and so on, with the condition that

$$t_1 + t_2 + \cdots + t_n = k.$$

Thus the coefficient of x^k will be precisely the number $f(n, k)$. It follows that

$$f(n, k) = \text{coefficient of } x^k \text{ in } (1.4)$$
$$= \text{coefficient of } x^k \text{ in } (1 + x + x^2 + \cdots)^n.$$

But

$$1 + x + x^2 + \cdots = \frac{1}{1 - x} = (1 - x)^{-1},$$

so that

$$f(n, k) = \text{coefficient of } x^k \text{ in } (1 - x)^{-n}. \quad (1.5)$$

This approach therefore reduces the problem to an application of the *binomial theorem*. This theorem will be studied in the next chapter, when it will be seen that

$$(1 - x)^{-n} = 1 + \frac{n}{1}x + \frac{n(n + 1)}{1.2}x^2 + \frac{n(n + 1)(n + 2)}{1.2.3}x^3 + \cdots$$

It will then follow that

$$f(n, k) = \frac{n(n + 1)(n + 2) \ldots (n + k - 1)}{1.2.3 \ldots\ldots k}$$

$$= \frac{n(n + 1) \ldots (n + k - 1)}{k!}$$

$$= \frac{(n + k - 1)!}{(n - 1)! \, k!}. \tag{1.6}$$

For example,

$$f(9, 4) = \frac{12!}{8! \, 4!} = \frac{12.11.10.9}{4.3.2.1} = 495.$$

The technique of picking out the coefficient of a particular power of x leads on to the general technique of *generating functions*. If a fixed value of n is chosen in the above example, then, by (1.5),

$$(1 - x)^{-n} = f(n, 0) + f(n, 1)x + f(n, 2)x^2 \ldots$$

since $f(n, k)$ is precisely the coefficient of x^k in the expansion of $(1 - x)^{-n}$. $(1 - x)^{-n}$ is called the generating function for the numbers $f(n, k)$ since its expansion generates these numbers as coefficients. Again, generating functions will be seen to be a useful tool.

Third approach

The problem is to colour k balls using n colours. One way of looking at this is to think of the colouring process as a splitting up of the k balls into n smaller collections some of which might be empty, where in the first collection x_1 balls are chosen and coloured with the first colour, and so on. For example, suppose that $n = 4$ and $k = 5$. One possible colouring would be to colour 2 balls with the first colour and one with each of the remaining colours. This corresponds to $x_1 = 2, x_2 = x_3 = x_4 = 1$, and can be represented geometrically as follows. Put two crosses at consecutive marks on a straight line. These crosses represent the two balls coloured with the first colour. At the next mark on the line place a O to signify a change of colour. Then put one cross to represent one ball of the second colour, then another O, and so on, to obtain

$$\times \times O \times O \times O \times$$

which has 5 $\times$s (the 5 balls) and 3 Os (three changes of colour). Similarly, corresponding to $x_1 = 2, x_2 = 0, x_3 = 2, x_4 = 1,$

$$\times \times O O \times \times O \times$$

would be obtained. In the general case, with n colours and k balls, $f(n, k)$ will be the number of such sequences of Os and $\times$s containing exactly k $\times$s and $(n - 1)$Os. Each such sequence contains $(n + k - 1)$ symbols altogether, and is uniquely determined once the Os have been positioned. Thus

$$f(n, k) = \text{number of ways of choosing } (n - 1) \text{ places out of } (n + k - 1) \text{ places.}$$

This third approach therefore leads to the following general question. Given m objects, how many ways are there of choosing exactly r of them? This is perhaps the most basic question in the whole subject, and so the next chapter will take a careful look at the problem of selections.

Exercises 1.2

1. Use (1.6) to evaluate $f(10, 3), f(10, 4), f(10, 5)$.

2. Verify that f, as given by (1.6), satisfies (1.1), (1.2), and (1.3).

3. By listing all the possibilities, show that there are 10 ways of choosing 3 girls from 5 given girls.

4. In a football league of n teams, each team plays each other twice. The number of games played is therefore $2c$ where c is the number of ways of choosing 2 objects from n given objects. Prove that

$$c = (n - 1) + (n - 2) + \cdots + 3 + 2 + 1 = \tfrac{1}{2}n(n - 1),$$

and deduce the number of games played in a league of 22 teams.

5. Let $h(n, k)$ denote the number of ways in which k indistinguishable golf balls can be coloured with n colours so that there is at least one ball of each colour. Prove that

   (a) $h(n, k) = 0$ if $n > k$,
   (b) $h(n, k)$ is the coefficient of x^k in $(x + x^2 + x^3 + \cdots)^n$,

(c) $h(n, k)$ is the coefficient of x^{k-n} in $(1-x)^{-n}$,

(d) $h(n, k) = f(n, k-n)$,

(e) $h(n, k) = \dfrac{(k-1)!}{(n-1)!\,(k-n)!}$ if $n > k$ (use (1.6) here).

Hence find $h(5, 10)$.

2 Selections and binomial coefficients

2.1. Permutations

THE basic problem here is: given n objects, how many ways are there of listing them in order? If $p(n)$ denotes the number of ways, then

$$p(1) = 1, \quad p(2) = 2, \quad p(3) = 6.$$

To see that $p(3) = 6$, let the objects be a, b, c. Then the 6 possible listings are

$$abc, \quad acb, \quad bac, \quad bca, \quad cab, \quad cba.$$

Any such ordering is called a *permutation* of a, b, c.

The general formula for $p(n)$ is now obtained. Choose one of the n objects to be placed first in the list. This can be done in n ways, and each of these n choices results in $(n - 1)$ objects being left. These $(n - 1)$ objects can be placed in the $(n - 1)$ remaining places in $p(n - 1)$ different orders, so that the recurrence relation

$$p(n) = np(n - 1)$$

is obtained. This, with the boundary condition $p(1) = 1$ gives, by Example 1.1,

$$p(n) = n!. \tag{2.1}$$

Example 2.1. A breakfast cereal competition lists 10 properties of a new make of car and asks the eater to place these properties in order of importance. (a) How many orderings are possible? (b) How many would be possible if the first and tenth places were already specified?

Solution. (a) $10!$; (b) 8 properties are left to be ordered. This can be done in $8!$ ways.

Example 2.2. A sports magazine decides to publish articles on all 22 first division (football) league clubs, one club per week for 22 weeks. In how many ways can this be done if the first article must be about Arsenal? How many if Wolves and Stoke must be featured on consecutive weeks?

Solution. (a) 21 teams are left to be ordered, so there are $21!$ orderings. (b) Consider Stoke-Wolves as one unit. Then this unit and 20 others have to be ordered. This can be done in 21! ways. But in each way there are two possible orderings of the Stoke-Wolves unit, so the required number is 2 x 21!

Exercises 2.1

1. How many 9-digit numbers can be obtained by using each of the digits 1, 2, ..., 9 exactly once? How many of these are bigger than 500 000 000?

2. How many permutations are there of the 26 letters of the alphabet in which the 5 vowels are in consecutive places?

3. The names of the 12 months of the year are listed in random order. Given that May and June are *not* next to each other, how many possible lists are there?

4. It is required to seat n people round a table. Show that this can be done in $(n - 1)!$ different ways. (Hint: put the n people in a row and then join up the ends of the row. Some rows will give the same circular arrangement.)

5. How many different necklaces can be designed from n colours, using one bead of each colour?

2.2. Ordered selections

The competition of Example 2.1 will now be slightly changed. Suppose the eater is now asked to choose only the 6 most important properties and to place these 6 in order of importance. How many possible lists are there now?

In general, let $p(n, r)$ denote the number of ways of listing r objects chosen from n. As for permutations above, the first object on the list can be chosen in n ways, and then $(r - 1)$ of the remaining $(n - 1)$ objects have to be added to the list. Thus

$$p(n, r) = np(n - 1, r - 1).$$

This gives

$$p(n, r) = n(n - 1) \ldots (n - r + 2)p(n - r + 1, 1),$$

where there is the boundary condition $p(s, 1) = s$ for all s. Thus

$$p(n, r) = n(n - 1) \ldots (n - r + 2)(n - r + 1)$$

$$= \frac{n!}{(n - r)!}. \tag{2.2}$$

In the above example, therefore, the number of possible lists is $\dfrac{10!}{4!}$.

Example 2.3. There are 5 seats in a row available, but 12 people to choose from. How many different seatings are possible?

Solution. $p(12, 5) = \dfrac{12!}{7!}$.

Example 2.4. 30 girls, including Miss U.K., enter a Miss World competition. The first 6 places are announced. (a) How many different announcements are possible? (b) How many if Miss U.K. is assured of a place in the first six?

Solution. (a) $p(30, 6) = \dfrac{30!}{24!}$;

(b) Here subtract from $p(30, 6)$ the number of placings which do *not* include Miss U.K. Such placings are in effect ordered selections of 6 from 29 candidates, so there are $p(29, 6)$ such orderings. The required number is therefore

$$\frac{30!}{24!} - \frac{29!}{23!} = \frac{29!}{24!}(30 - 24) = \frac{6 \cdot 29!}{24!}.$$

In the above examples, once an object has been chosen it cannot be chosen again. However, sometimes repetitions are allowed.

Example 2.5. For each day of the 5-day working week I can choose any one of 4 newspapers to read in the train. How many different buys are possible in a week?

Solution. The point here is that if I buy the Times on Monday, I can still buy the Times later on in the week. For each day there are 4 choices, so the total number of choices for the week is

$$4 \times 4 \times 4 \times 4 \times 4 = 4^5 = 1024.$$

Exercises 2.2

1. Evaluate $p(7, 4)$, $p(8, 2)$, $p(9, 5)$.

2. A car registration number is to consist of 3 letters followed by a number between 1 and 999. How many car numbers are possible?

3. Tom has 75 books but enough room on his bookshelf for only 20. In how many ways can he fill his shelf?

4. How many numbers between 1000 and 3000 can be formed from the digits 1, 2, 3, 4, 5 if repetition of digits is (a) allowed, (b) not allowed?

5. In twelve-tone music, the twelve notes of the chromatic scale are put in a row, and have to be played in that particular order. How many rows are possible?

6. There are k stations on a particular railway. If it is possible to travel from any station to any other, how many different journeys are possible?

7. In how many ways can a 5-letter word be formed from an alphabet of 26 letters if repetitions are (a) allowed, (b) not allowed?

8. A *binary sequence of length n* is a string of n digits each of which is 0 or 1. How many such sequences are there? List all those of length 4.

2.3. Unordered selections

Often in making a selection, the selected objects are not placed in any particular order. For example, if 5 out of 8 books are to be chosen, the only interest is in which 5 are chosen, not in the order in which they are chosen. How many ways are there of choosing 5 books from 8? More generally, let $c(n, r)$ denote the number of ways of choosing r objects from n given objects, without taking order into account. Consider any selection of r objects. This selection can be ordered in $p(r) = r!$ different ways, and so each unordered selection gives rise to $r!$ ordered selections. Thus

$$r!\, c(n, r) = \text{total number of ordered selections}$$

$$= \frac{n!}{(n - r)!}$$

so that

$$c(n,r) = \frac{n!}{r!\,(n - r)!}. \tag{2.3}$$

This number is often written as $\binom{n}{r}$. Thus

$$\binom{n}{r} = \frac{n!}{r!\,(n - r)!}.$$

For example,

$$c(8, 5) = \binom{8}{5} = \frac{8!}{5!\,3!} = \frac{8.7.6}{3.2.1} = 56.$$

There are therefore 56 ways of choosing 5 books from 8.

Example 2.6. In Chapter 1, the third method of attacking the problem gave

$$f(n, k) = c(n + k - 1, n - 1).$$

Thus

$$f(n, k) = \binom{n + k - 1}{n - 1} = \frac{(n + k - 1)!}{(n - 1)! \, (n + k - 1 - n + 1)!} = \frac{(n + k - 1)!}{(n - 1)! \, k!},$$

agreeing with method 2.

One of the important properties of the numbers $\binom{n}{r}$ is given in the following theorem. The convention that $\binom{n}{0} = 1$ is followed.

THEOREM 2.1. $\binom{n}{r} = \binom{n}{n - r}$. $(0 \leqslant r \leqslant n)$.

Proof. Two alternative proofs are given, both of which should be studied.

First proof
Selecting r objects from n is equivalent to choosing the $(n - r)$ objects which shall not be selected!

Second proof

$$\binom{n}{r} = \frac{n!}{r! \, (n - r)!} = \frac{n!}{(n - (n - r))! \, (n - r)!} = \binom{n}{n - r}.$$

Example 2.7. (1) $\binom{8}{3} = \binom{8}{5}$. (2) $\binom{n}{n - 1} = \binom{n}{1} = n$ and

$\binom{n}{n - 2} = \binom{n}{2} = \frac{1}{2}n(n - 1)$ for all n.

The numbers $\binom{n}{r}$ are of extreme importance in mathematics. This is because of the following theorem.

THEOREM 2.2. *Let n be a positive integer. Then, if $(1 + x)^n$ is expanded as a sum of powers of x, the coefficient of x^r is $\binom{n}{r}$.*

Example 2.8.

$$(1 + x)^0 = 1$$
$$(1 + x)^1 = 1 + x$$
$$(1 + x)^2 = 1 + 2x + x^2$$
$$(1 + x)^3 = 1 + 3x + 3x^2 + x^3$$
$$(1 + x)^4 = 1 + 4x + 6x^2 + 4x^3 + x^4$$
$$(1 + x)^5 = 1 + 5x + 10x^2 + 10x^3 + 5x^4 + x^5.$$

Proof of Theorem 2.2. Consider the product

$$(1 + x)(1 + x) \ldots (1 + x) \quad (n \text{ brackets}).$$

A term x^r is obtained by choosing r of the brackets, selecting the term x from each of them, and selecting the term 1 from the remaining $(n - r)$ brackets. Thus the number of times x^r is obtained is just the number of ways of choosing r of the n brackets, i.e. $c(n, r)$.

The coefficients in the expansions are, by Example 2.8,

$$
\begin{array}{c}
1 \\
1 \quad 1 \\
1 \quad 2 \quad 1 \\
1 \quad 3 \quad 3 \quad 1 \\
1 \quad 4 \quad 6 \quad 4 \quad 1 \\
1 \quad 5 \quad 10 \quad 10 \quad 5 \quad 1 \\
\vdots \qquad\qquad \vdots
\end{array}
$$

This array is known as *Pascal's triangle*. The $(n + 1)$th row gives the

numbers $\binom{n}{0}, \binom{n}{1}, \ldots, \binom{n}{n}$. The property of Theorem 2.1 is simply that

each row reads the same forwards as backwards. But another property is clear in the triangle: each number in the array is the sum of the two numbers immediately above it. This is because of the following recurrence relation, for which again two proofs are given.

THEOREM 2.3. $\binom{n}{r} = \binom{n - 1}{r - 1} + \binom{n - 1}{r}.$

First proof. $\binom{n}{r}$ is the number of ways of choosing r objects from n. Any

particular choice may or may not include the nth object. If the nth object is included, the problem is that of choosing $(r-1)$ from the remaining $(n-1)$, and this can be done in $\binom{n-1}{r-1}$ ways. If the nth object is not chosen, r objects have to be selected from the remaining $(n-1)$, and this can be done in $\binom{n-1}{r}$ ways.

Second proof. $\binom{n-1}{r-1} + \binom{n-1}{r} = \dfrac{(n-1)!}{(r-1)!\,(n-r)!} + \dfrac{(n-1)!}{r!\,(n-r-1)!}$

$$= \frac{(n-1)!}{r!\,(n-r)!}\{r+(n-r)\} = \frac{n!}{r!\,(n-r)!} = \binom{n}{r}.$$

Example 2.9. $\binom{7}{4} = \binom{6}{3} + \binom{6}{4}.$

Theorem 2.2 can be re-expressed in the following form.

THEOREM 2.2*. *If n is any positive integer, then*

$$(1+x)^n = \binom{n}{0} + \binom{n}{1}x + \binom{n}{2}x^2 + \cdots + \binom{n}{n}x^n = \sum_{r=0}^{n}\binom{n}{r}x^r.$$

More generally, the following results holds.

THEOREM 2.4. *If n is any positive integer, then*

$$(a+b)^n = \binom{n}{0}a^n + \binom{n}{1}a^{n-1}b + \binom{n}{2}a^{n-2}b^2 + \cdots + \binom{n}{n}b^n$$

$$= \sum_{r=0}^{n}\binom{n}{r}a^{n-r}b^r.$$

Proof. $(a+b)^n = a^n\left(1 + \dfrac{b}{a}\right)^n = a^n\sum_{r=0}^{n}\binom{n}{r}\left(\dfrac{b}{a}\right)^r = \sum_{r=0}^{n}\binom{n}{r}b^r a^{n-r}.$

Theorem 2.4 is known as the *binomial theorem*, and the numbers $\binom{n}{r}$ are called the *binomial coefficients*. The name 'binomial' refers to the fact that the theorem is concerned with the expansion of the nth power of a sum of *two* symbols. As an example of the theorem,

$$(x + y)^7 = x^7 + 7x^6y + 21x^5y^2 + 35x^4y^3 + 35x^3y^4 + 21x^2y^5 + 7xy^6 + y^7.$$

In the language of Chapter 1, $(1 + x)^n$ is the *generating function* of the binomial coefficients. One useful special case is the following.

THEOREM 2.5. *If n is any positive integer, then*

$$(1 - x)^n = 1 - \binom{n}{1}x + \binom{n}{2}x^2 - \cdots + (-1)^n\binom{n}{n}x^n = \sum_{r=0}^{n}\binom{n}{r}(-1)^r x^r.$$

This follows from the binomial theorem on choosing $a = 1$, $b = -x$.

Example 2.10. Later on in this book, it will be necessary to consider the following series:

$$\exp(x) = 1 + x + \frac{x^2}{2!} + \frac{x^3}{3!} + \frac{x^4}{4!} + \cdots.$$

Using the binomial theorem, it is easy to prove the following important property:

$$\exp(x)\exp(y) = \exp(x + y).$$

For the left-hand side is

$$\left(1 + x + \frac{x^2}{2!} + \frac{x^3}{3!} + \cdots\right)\left(1 + y + \frac{y^2}{2!} + \frac{y^3}{3!} + \cdots\right),$$

so that, when the brackets are multiplied together, the terms of the form $x^r y^s$ with $r + s = n$ which are obtained are precisely

$$\frac{x^n}{n!} + \frac{x^{n-1}}{(n-1)!}\frac{y}{1!} + \frac{x^{n-2}}{(n-2)!}\frac{y^2}{2!} + \cdots + \frac{x}{1!}\frac{y^{n-1}}{(n-1)!} + \frac{y^n}{n!}$$

$$= \frac{1}{n!}\left\{x^n + \frac{n!}{(n-1)!\,1!}x^{n-1}y + \frac{n!}{(n-2)!\,2!}x^{n-2}y^2 + \cdots + y^n\right)$$

$$= \frac{1}{n!}(x + y)^n.$$

Already in this book the need for an expansion for $(1 - x)^{-n}$ has been met. Such an expansion is impossible if it is required that there should be only a finite number of terms, as happens in the expansion of $(1 - x)^n$, but an infinite series representing $(1-x)^{-n}$ can be obtained. In fact, since the problem which gave rise to $(1-x)^{-n}$ has already been solved by another method (method 3), we can turn this to our advantage and use it to prove:

THEOREM 2.6. *If n is any positive integer, then*

$$(1-x)^{-n} = 1 + \binom{n}{1}x + \binom{n+1}{2}x^2 + \binom{n+2}{3}x^3 + \cdots$$

$$= \sum_{r=0}^{\infty} \binom{n+r-1}{r}x^r.$$

Proof. The theorem simply states that $f(n, r) = \binom{n+r-1}{r}$. But this has been proved by method 3 (see Example 2.6).

Example 2.11. $(1-x)^{-4} = 1 + 4x + 10x^2 + 20x^3 + \cdots$

Example 2.12. Use Pascal's triangle and the fact that

$$f(n, r) = \binom{n+r-1}{r}$$

to extend the following table of values of $f(n, r)$.

r＼n	1	2	3	4	5	6
1	1	2	3	4		
2	1	3	6	10		
3	1	4	10	20		
4						

Exercises 2.3

1. Expand $(1+x)^8$ and $(1-x)^8$.

2. Evaluate $\binom{11}{4}$, $\binom{13}{7}$, $\binom{15}{8}$.

3. Obtain the first few terms in the expansion of $(1-x)^{-8}$.

4. How many solutions are there of the equation $x + y + z = 10$ with x, y, z non-negative integers?

5. How many solutions are there of the equation $x + y + z = 10$ with x, y, z positive integers?

6. An eight-man committee is to be formed from a group of 10 Welshmen and 15 Englishmen. In how many ways can the committee be chosen if
   (a) the committee must contain 4 of each nationality,
   (b) there must be more Welshmen than Englishmen,
   (c) there must be at least two Welshmen?

7. A king is placed on the bottom left hand square of an 8 x 8 chessboard and is to move to the top right-hand corner square. If it can move only up or to the right, how many possible paths does it have to choose from?

8. By using the identity

$$(1 + x)^{2n} = (1 + x)^n (1 + x)^n,$$

and considering the coefficient of x^n on both sides, prove that

$$\binom{2n}{n} = \binom{n}{0}^2 + \binom{n}{1}^2 + \binom{n}{2}^2 + \cdots + \binom{n}{n}^2.$$

Verify this in the case $n = 5$.

9. Prove that $\binom{n+1}{3} - \binom{n-1}{3} = (n-1)^2$.

10. Show that, in each row of Pascal's triangle, the largest number is the middle one (or two), i.e. show that

$$\binom{n}{r} > \binom{n}{r-1} \quad \text{if } r < \tfrac{1}{2}(n+1).$$

11. A team of 11 players is to be chosen from a pool of 15. How many team selections are possible? How many if one of the 15 has already been appointed captain, and must play?

12. Prove that

$$\binom{n}{0} + \binom{n+1}{1} + \binom{n+2}{2} + \cdots + \binom{n+m}{m} = \binom{n+m+1}{m}.$$

13. Suppose that $r < m$, and consider the identity
$$(1 - x)^m (1 - x)^{-(r+1)} = (1 - x)^{m-r-1}.$$

Expand the brackets on the left using Theorems 2.5 and 2.6. The coefficient of x^{m-r} on the left must be 0 (why?). Show that this coefficient is

$$\binom{m}{m-r}\binom{m}{0} - \binom{m-1}{m-r-1}\binom{m}{1} + \binom{m-2}{m-r-2}\binom{m}{2} - \cdots$$

and deduce that

$$\sum_{s=r}^{m} \binom{m}{s}\binom{s}{r}(-1)^{m-s} = 0$$

if $r < m$.
Verify this result when $r = 6$, $m = 9$.

2.4 Further remarks on the binomial theorem

The binomial theorem can be written as

$$(1+x)^m = 1 + mx + \frac{m(m-1)}{2!}x^2 + \frac{m(m-1)(m-2)}{3!}x^3 + \cdots \tag{2.4}$$

where the series on the right terminates after $(m+1)$ terms if m is a positive integer, but continues indefinitely if m is a negative integer. The final remark in this chapter is that, although it will not be proved in this book, the expansion (2.4) also holds if m is not an integer. In particular, $m = \frac{1}{2}$ gives

$$(1+x)^{1/2} = 1 + \tfrac{1}{2}x + \tfrac{1}{2}.(-\tfrac{1}{2})\frac{x^2}{2!} + \tfrac{1}{2}.(-\tfrac{1}{2}).(-\tfrac{3}{2})\frac{x^3}{3!} + \cdots$$

where, for $k \geqslant 2$, the coefficient of x^k is

$$(-1)^{k+1}\frac{1.3.5\ldots(2k-3)}{2^k.k!} = \frac{(-1)^{k+1}(2k-2)!}{2^k k!\, 2.4\ldots(2k-2)}$$

$$= \frac{(-1)^{k+1}(2k-2)!}{2^k k!\, 2^{k-1}(k-1)!} = \frac{(-1)^{k+1}(2k-2)!}{2^{2k-1}k!\,(k-1)!}$$

This expansion of $(1+x)^{1/2}$ will be of use when generating functions are discussed in Chapter 4. For a proof of the validity of the expansion, the reader is referred to almost any introductory calculus book.

Exercises 2.4

1. Write down the first few terms of the expansion of $(1 - x)^{-1/2}$, and show that the coefficient of x^k in it is

$$\frac{1}{4^k}\binom{2k}{k}.$$

2. Write down the first few terms of the expansions of $(1 + x)^{1/4}$, $(1 - x)^{1/4}$, $(1 + x)^{-1/4}$, $(1 - x)^{-1/4}$.

3. Using the identity $(1 - x)^{-1} = (1 - x)^{-1/2}(1 - x)^{-1/2}$, and the expansion obtained in Question 1 above, and equating the coefficients of x^k on both sides of the identity, show that

$$\binom{2k}{k} + \binom{2k-2}{k-1}\binom{2}{1} + \binom{2k-4}{k-2}\binom{4}{2} + \cdots + \binom{2k}{k} = 4^k.$$

2.5. Miscellaneous problems on Chapter 2

1. In how many ways can four tyres be put on a car?

2. 16 people, 4 from each of Scotland, England, Ireland, and Wales, have to select 6 of their number to represent them on a committee. How many selections can be made if
   (a) each country must be represented?
   (b) no country can have more than two representatives?

3. How many solutions are there in non-negative integers of
   (a) $x + y + z = 8$?
   (b) $x + y + z + w = 18$?

4. I have time to visit one friend on each evening of a given week. There are 12 friends whom I would like to visit. In how many ways can I plan my week if
   (a) I can visit a friend more than once?
   (b) I will not visit a friend more than once?

5. Prove that $\binom{n}{r}^2 > \binom{n}{r-1}\binom{n}{r+1}$ for all $n > r$.

6. Six men are to be seated round a circular table. How many ways are there of achieving this? How many if A refuses to sit beside B?

7. Prove that $\binom{n}{0} + \binom{n}{1} + \binom{n}{2} + \cdots + \binom{n}{n} = 2^n$. (Use Theorem 2.2*

   with x suitably chosen.)

8. Similarly prove that $\binom{n}{0} - \binom{n}{1} + \binom{n}{2} - \cdots + (-1)^n \binom{n}{n} = 0$.

9. Let A be a set with n elements. Show that the number of subsets of A with r elements is $\binom{n}{r}$. Show also that the number of subsets is 2^n.

   (Hint: consider each element of A in turn—it can be either in or not in any given subset—two possibilities.) Deduce the result of Question 7.

10. A cube, with each face a different colour, fits perfectly into a cubical box. How many ways are there of putting the cube into the box?

3 Pairings problems

3.1. Pairings within a set

PAIRINGS problems fall, roughly speaking, into two categories. The first type is concerned with splitting up a set with an even number of elements into pairs of elements, for example arranging $2n$ students in n pairs to share rooms in a college residence hall. The second type is concerned with pairing off the elements of one set with those of another, for example assigning jobs to applicants so that no two applicants get the same job. A problem of the first type will serve as the starting point of this chapter. Given $2n$ objects, how many ways are there of forming n pairs?

> **Example 3.1.** Six men A, B, C, D, E, F are to be paired off. One way is A with B, C with D, E with F, whereas another way is A with C, B with F, E with D. There are 15 possible ways altogether and the reader is left to produce the remaining 13. This method is rather lengthy, and so a better method is looked for.

In general, when there are $2n$ objects, a first idea might be to place these objects in n brackets (2 in each) strung in a row as shown.

$$(\ , \ )(\ , \ ) \ldots (\ , \ )$$

The objects can be placed in the spaces in $(2n)!$ different ways. In each bracket, however, there are 2! different orderings, which have to be considered as giving the same pairing, so the number $(2n)!$ must be divided by 2! for each bracket, i.e. by $(2!)^n$. Further, the order of the brackets does not matter, and since the brackets can be arranged in $n!$ different ways, each distinct pairing has in fact been obtained $n!$ times. On dividing by $n!$ the total number of different pairings is finally

$$\frac{(2n)!}{(2!)^n n!} \tag{3.1}$$

For example $n = 3$ gives $\dfrac{6!}{2^3 3!} = 15$ as already observed. This method clearly generalizes to the problem of splitting up mn objects into n sets of m objects, the working above simply corresponding to $m = 2$.

THEOREM 3.1. *Let S be a set of mn objects. Then S can be split up (partitioned) into n sets of m elements in*

$$\frac{(mn)!}{(m!)^n n!}$$

different ways.

Proof. Replace 2 by m in the above argument.

Example 3.2. A wholesale company has to supervise sales in 20 towns. Five members of staff are available, and each is to be assigned 4 towns to supervise.

(a) In how many ways can the 20 towns be put into 5 groups of 4?

(b) In how many ways can the towns be assigned to the staff?

Solution. (a) The theorem gives the number as $\frac{20!}{(4!)^5 5!}$. (b) Imagine that the towns have been arranged in 5 groups of 4 in some particular way. Then the 5 groups can be assigned to the 5 men in 5! different ways, depending on which group goes to the first man, which to the second, and so on. The required number is therefore 5! times the number in part

(a), i.e. $\frac{20!}{(4!)^5}$.

Note on (a). This corresponds to omitting the last part of the argument which proved (3.1). Here the order of the brackets does matter, since the first bracket corresponds to the first man, and so on.

In practice, of course, things are generally far more complicated. In Example 3.1, for example, A may refuse to be paired with B. This gives a new problem. Instead of asking how many pairings are possible, the question becomes: does even one pairing exist, taking into account the likes and dislikes of the six people?

Example 3.3. In Fig. 3.1, the 6 dots represent 6 people. Two dots are joined by a line if and only if the two people represented by the dots are willing to be paired together. Is it possible to achieve a pairing?

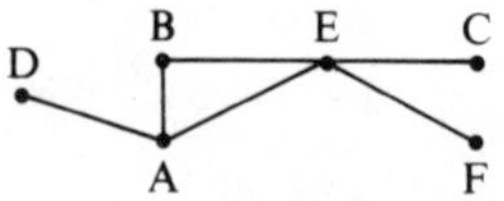

FIG. 3.1

Solution. No. For C can be paired only with E, and this leaves no-one to be paired with F.

A diagram such as Fig. 3.1 is called a *graph*. The dots are known as the *vertices* and the lines as the *edges*. The convention will be followed in this book that any pair of vertices of a graph can be joined by at most one edge.

Roughly speaking, the failure to find a pairing in the above example is due to the lack of edges. Vertex F has only one edge emanating from it, whereas it could have as many as five. The following theorem shows that if each vertex has at least half the possible edges from it present, then a pairing can be achieved. The proof is constructive in the sense that it not only proves there is a pairing but it describes how a pairing can be found in practice by a routine procedure. Such a procedure is called an *algorithm*, and can be programmed for a computer.

Two definitions are given before the proof. The *degree* of a vertex of a graph is the number of edges with that vertex as an end-point. For example, in Fig. 3.1, the respective degrees are 3, 2, 1, 1, 4, 1. Also, a pairing off of all the vertices of a graph is often called a *complete* or *perfect matching* of the graph. Clearly a graph needs to have an even number of vertices if it is to have a perfect matching.

THEOREM 3.2. *If a graph has 2n vertices, each of degree $\geqslant n$, then the graph has a perfect matching.*

Proof. Assuming that r pairs of vertices have so far been paired off, where $r < n$, the proof shows how to increase this to $(r + 1)$ pairs. If there are two vertices not yet paired off but joined by an edge, they can be taken immediately as the $(r + 1)$th pair. So suppose now that no two of the remaining vertices are joined by an edge. Choose any two of them, and call them a and b. It will now be shown that there must be a pair u, v of vertices already paired together such that a and u are joined by an edge and b and v are joined by an edge (see Fig. 3.2). The pairings can then be rearranged so that a is paired with u and b with v, thus increasing the number of pairs to $(r + 1)$.

FIG. 3.2

Suppose no pair u, v exists. Then each of the r pairs x, y of vertices so far formed is such that at most two of the four possible edges ax, ay, bx, by actually appear in the graph. Thus the total number of edges from the r pairs to a and b is at most $2r < 2n$. But since a and b are both of degree $\geqslant n$, the number must be $\geqslant 2n$, giving a contradiction.

The algorithm is therefore as follows. Having obtained r pairs, scan the remaining $(2n - 2r)$ vertices to see if two of them are joined by an edge. If not, choose any two of them, a, b, and scan the pairs x, y already formed until one is found such that a is joined to x and b to y. Then replace the pair x, y by the two pairs a, x and b, y. If $r + 1 < n$, repeat the whole process.

Exercises 3.1

1. 10 people meet and form 5 pairs. In how many ways can these 5 pairs be formed?

2. 16 teams qualify for a particular round of the F.A. Cup. How many possible pairings are there for the 8 games if it (a) is (b) is not taken into account which teams are drawn at home?

3. A pack of 52 cards is divided among 4 people so that each gets 13 cards (as in bridge). How many such deals are possible?

4. In the Scottish League Cup, 16 first division clubs are arranged in 4 groups of 4. In how many ways can this be done? Recently, Rangers and Celtic were drawn in the same section two years running. Show that this is not as strange as the press made it out to be by finding the number of ways the draw can be made with Rangers and Celtic in the same section, and verifying that this number is precisely one fifth of the total number of possible draws.

5. The following are all the allowable pairings of 8 objects. $(1, 2)$, $(1, 3)$, $(2, 4)$, $(2, 5)$, $(4, 6)$, $(5, 7)$, $(5, 8)$, $(7, 8)$. Obtain a complete matching.

6. Draw a graph with 10 vertices, each of degree $\geqslant 5$, and find a perfect matching for it.

7. Construct a graph with 10 vertices, each of degree $\geqslant 4$, with no perfect matching.

3.2. Pairings between sets

A number of jobs are available in a large industrial organization, and applicants are examined for suitability for each of the jobs. In what circumstances is it possible to assign a suitable person to each?

This problem, one type of *assignment problem*, is typical of those to be examined in this section. More generally, given two sets A, B (here, the set of jobs and the set of applicants), when is it possible to pair off each member of A with a different member of B?

Example 3.4. Five jobs are available. For each $i = 1, \ldots, 5$, let S_i denote the set of applicants suited for the ith job. Can all the jobs be filled?

$$S_1 = \{A, B, C\}, \; S_2 = \{D, E\}, \; S_3 = \{D\},$$
$$S_4 = \{E\}, \; S_5 = \{A, E\}.$$

Solution. No. The second, third, and fourth jobs have only 2 suitable applicants between them. But 2 men cannot fill 3 jobs.

This example deserves closer scrutiny. By introducing the sets $S_1, \ldots,$ S_5, the problem has been re-expressed as one of the following type. Given sets $S_1, \ldots, S_n$, is it possible to choose a different element from each set S_i? If it is possible, then the chosen elements are called *distinct representatives* of the sets. In the above example, the sets S_3, S_4, S_5 possess distinct representatives (D, E, and A, in that order), but the sets S_1, S_2, S_3, S_4, S_5 do not. The reason is:

There are 3 sets containing between them less than 3 elements.

Clearly, if distinct representatives do exist, then, for every value of k;

any k sets contain between them at least k elements. (3.2)

This is a *necessary* condition. The interesting and useful fact is that the condition is not only necessary but it is also *sufficient*. In other words, if (3.2) holds for every value of k, then it is guaranteed that distinct representatives can be found. The proof which will be given is an algorithm which not only shows that distinct representatives exist, but gives a method of actually finding them. The following result will then have been proved.

THEOREM 3.3. (Philip Hall's theorem on distinct representatives). *The sets $A_1, \ldots, A_n$ possess a system of distinct representatives if and only if, for all $k = 1, \ldots, n$, any k A_is contain at least k elements in their union.*
An alternative formulation would be:

ASSIGNMENT THEOREM. *This assignment problem has a solution if and only if there is no value of k for which there are k jobs with fewer than k suitable applicants between them.*
Replacing the job situation by marriage gives yet another formulation

of Hall's theorem which has earned it the popular title of the *Marriage Problem*.

MARRIAGE THEOREM . *Given a set of men and a set of women, each man makes a list of the women he is willing to marry. Then each man can be married off to a woman on his list if and only if,*

$$(*) \quad \begin{cases} \textit{for every value of } k, \textit{ any } k \textit{ lists contain in} \\ \textit{their union at least } k \textit{ names.} \end{cases}$$

Proof of the Marriage Theorem. It is shown how, on the assumption that $r < n$ men have been paired off with suitable ladies, to increase this to $(r + 1)$ men.

Suppose r men have been paired off. If there is a man left who has on his list a woman who is still unattached, an $(r + 1)$th pairing is immediate. So suppose that all women on remaining lists are already attached.

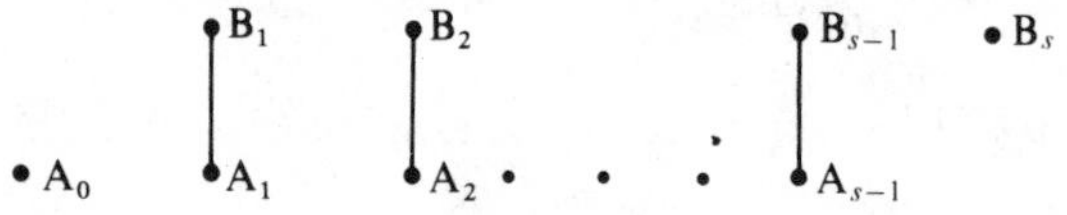

FIG. 3.3

Choose any unmarried man A_0 (see Fig. 3.3). By (*) with $k = 1$, there is a woman B_1 on his list. B_1 is married to A_1, say. By (*) with $k = 2$, the combined lists of A_0 and A_1 contain the name of at least one more woman B_2. If B_2 is unmarried, stop. If B_2 is married to A_2, then, by (*) with $k = 3$, the combined lists of A_0, A_1, A_2 contain a third name, say B_3. If B_3 is unmarried, stop. If B_3 is married to A_3, repeat the process, and continue until an unmarried woman B_s is reached. (This must happen eventually since not all the women are married, and no B_i occurs twice in the process.)

Note that, by construction, each B_i is on the list of at least one A_j with $j < i$. This is very important. Consider now B_s. Pair her off with an A_i on whose list she appears $(i < s)$. This frees B_i. Next pair off B_i with an A_j $(j < i)$ on whose list she appears. This frees B_j. Repeat until some B is freed and re-paired with A_0. This must eventually happen. Then take all the new pairings and all the original ones which have not been tampered with. Now $(r + 1)$ pairs have been obtained. Repeat the process if $r + 1 < n$.

This constructive proof, which seems to have been presented independently by R. Rado, D. J. Shoesmith, and others, has the advantage that the conditions (*) need not be checked before the construction is attempted. If (*) does not hold, this will become clear when the method breaks down. On the other hand, if (*) holds, the method will not break down.

Application to Latin squares

The three squares

1	2
2	1

1	3	2
2	1	3
3	2	1

1	2	3	4
3	4	1	2
2	1	4	3
4	3	2	1

each possess the following properties (for n = 2, 3, 4 respectively),
   (1) each row contains each of the numbers 1, . . ., n exactly once;
   (2) each column contains each of 1, . . ., n exactly once.
These are the properties which characterize *Latin squares*. An $n \times n$ Latin square based on the numbers 1, . . ., n is thus defined to be an array of n rows and n columns satisfying properties (1) and (2) above. Apart from their intrinsic mathematical charm, Latin squares do have their uses, and may first have been studied because of their uses in the design of experiments. A simple introduction to this topic can be found in Fisher's classical book [1].

   How are Latin squares constructed? Very easily, in fact, for the next result shows that the construction can be carried out a row at a time. The proof requires the idea of a *Latin rectangle*, which is simply a rectangular array with r rows and n columns ($r \leqslant n$) in which
   (1) each row contains each of 1, . . ., n exactly once;
   (3) no column contains a number more than once.
For example, the first three rows of the 4 x 4 Latin square above give a 3 x 4 Latin rectangle. The theorem to be proved is essentially a converse of this result: it states that any Latin rectangle can be made into a Latin square by adding further rows, without having to alter the rows already there.

   THEOREM 3.4. *If* $r < n$, *any* $r \times n$ *Latin rectangle can be extended to an* $(r + 1) \times n$ *Latin rectangle.*

Proof. An $(r + 1)$th row has to be added, the jth number in which does not yet occur in the jth column of the rectangle. This suggests that for each $j = 1, \ldots, n$ the set S_j should be defined as follows:

S_j = set of numbers between 1 and n which have not yet appeared in the jth column.

To prove the theorem it is sufficient to show that the sets S_j possess distinct representatives. These distinct representatives will form the next row.

Suppose then that the sets S_j do not possess distinct representatives. Then by Hall's theorem, there must, for some k, be k sets S_i which in their union contain less than k numbers. Now clearly each set S_i has $(n - r)$ elements, so these k sets contain between them $k(n - r)$ numbers, not taking repetitions into account. How many times can a number be repeated? Each number has occurred exactly once in each row, and hence in exactly r of the columns. Each number therefore occurs in exactly $(n - r)$ of the sets S_i. The k sets therefore contain $k(n - r)$ elements, with no element repeated more than $(n - r)$ times, and so must contain at least k distinct elements. This gives a contradiction, and the proof is complete.

Exercises 3.2

1. If $A_1 = \{1, 2\}, A_2 = \{4\}, A_3 = \{1, 3\}$, and $A_4 = \{2, 3, 4\}$, find distinct representatives for the sets A_i.

2. Find a set of distinct representatives for the following sets: $\{a\}$, $\{a, b, c\}$, $\{c, d\}$, $\{b, d, e\}$, $\{c, f\}$, $\{a, d, g\}$, $\{f\}$.

3. Construct 2 different 5 x 5 Latin squares which have the same first rows, but no other rows the same.

4. S is a set of mn elements, and it is partitioned into n sets of m elements in two different ways:

$$S = A_1 \cup \cdots \cup A_n = B_1 \cup \cdots \cup B_n.$$

By considering the sets C_i defined by

$$C_i = \text{set of } j\text{s such that } B_j \text{ intersects } A_i,$$

show that the sets $C_1, \ldots, C_n$ have distinct representatives and deduce that the Bs can be relabelled so that $A_i \cap B_i$ is non-empty for each i.

5. *mn* newspaper reporters each cover one sport and one foreign country, in such a way that each of *n* sports has *m* reporters and each of *n* countries has *m* reporters. Use the previous example to show that it is possible to staff *m* newspapers each with *n* reporters so that each sport and each country is covered by each newspaper.

3.3. An optimal assignment problem

In the assignment problem considered in §3.2, each applicant was simply either suitable or not suitable for a particular job; there was no hint given as to the relative suitabilities of the candidates. The problem now to be considered arises when there are *n* candidates for *n* jobs, each candidate with a measure of his suitability for each of the jobs, and it is required to assign the candidates to the jobs in the best possible way. In general it will not be possible to assign each candidate to the job he is best suited for, since two candidates may both be best suited for the same job, and so an overall optimal assignment is to be looked for.

As a particular example, suppose there are four jobs *a*, *b*, *c*, *d* and four applicants *A*, *B*, *C*, *D*. For reasons of convenience, a table is constructed giving a measure of unsuitability, rather than suitability. Such a table may be as follows.

	A	B	C	D
a	5	7	15	12
b	8	3	9	10
c	4	14	2	5
d	6	3	1	14

If *A* is given the job *a*, *B* given *b*, *C* given *c* and *D* given *d*, the total measure of unsuitability is 5 + 3 + 2 + 14 = 24, whereas if *A* is given *a*, *B* given *b*, *C* given *d* and *D* given *c*, the total measure of unsuitability is only 14, and this assignment is obviously better than the first. The problem of finding the best possible assignment is clearly equivalent to the problem of selecting four numbers from the table, no two in the same row or column, with the smallest possible sum. In solving this problem it is important to notice the fact that the problem is unchanged if the same number is subtracted from all the members of any row or column, for exactly one number will be selected from that row or column and so all sums corresponding to assignments will be reduced by the same amount, thus leaving the best assignment as still the best. This means that the numbers in the table can be made easier to deal with by first of all subtracting from each row the smallest number in that row; all the resulting numbers will be non-negative and smaller than before. The table above becomes

	A	B	C	D
a	0	2	10	7
b	5	0	6	7
c	2	12	0	3
d	5	2	0	13.

(3.3)

Now subtract from each column the smallest number in that colum to obtain

	A	B	C	D
a	0	2	10	4
b	5	0	6	4
c	2	12	0	0
d	5	2	0	10.

The problem is still to select four numbers from this table, no two in the same row or column, with the smallest possible sum. Since all the numbers are non-negative, the smallest sum which can possibly occur is 0. Therefore, if four 0s can be found, no two in the same row or column, the places in which they occur will give a best possible assignment. In the table above, four such 0s can be found, in the places corresponding to *Aa, Bb, Cd, Dc*. This is therefore the best possible assignment of jobs to the candidates.

　　It was fortunate in this example that at the final stage four 0s were present in different rows and columns. In general, this will not happen, and the problem is then what to do next. At this stage a very important fact must be pointed out. It is a consequence of Hall's theorem on distinct representatives that the largest number of 0s which can be chosen from an $n \times n$ array, no two in the same row or column (*independent* 0s, as they are called) is always equal to the smallest number of rows and columns which together contain all the 0s. For example, in (3.3) three independent 0s can be found, whereas all the 0s can be covered by three lines, namely the first three columns or, alternatively, the first two rows and the third column. This result gives a method of checking whether or not n independent 0s can be found. For if all the 0s can be included in less than n lines, then n independent 0s cannot possibly be found. However, this result has another application. For if it happens that all the 0s can be included in less than n lines, another 0 can be produced elsewhere in the table by subtracting a suitable number from one of the other rows or columns. In general this will result in more lines being required to include all the 0s, and this in turn, by the result quoted, will mean that there are more independent 0s in the array.

The procedure is illustrated in the following example. Suppose the unsuitability table is

6	8	2	7
5	8	13	9
2	7	8	9
4	11	7	10.

Subtracting the smallest number from each row in turn gives

4	6	0	5
0	3	8	4
0	5	6	7
0	7	3	6.

Now subtract the smallest number from each column in turn to get

4	3	0	1
0	0	8	0
0	2	6	3
0	4	3	2.

There are not four independent 0s since three lines, namely the first column and the first two rows, include all the 0s. The next step is the following. Let m denote the smallest number which is not included in these lines. Here $m = 2$. Then subtract m from all the uncrossed columns, and add m to all the crossed rows. The reader should check that the net effect of this is to

(a) subtract m from all uncrossed numbers,
(b) leave numbers which are crossed once unchanged,
(c) add m to all numbers which are crossed twice.

This produces at least one more 0 in an uncrossed position, and leaves all the 0s already present unchanged, unless they happen to lie in both a crossed out row and a crossed out column. In the example above, the array becomes

$$
\begin{array}{cccc}
6 & 3 & 0 & 1 \\
2 & 0 & 8 & 0 \\
0 & 0 & 4 & 1 \\
0 & 2 & 1 & 0.
\end{array}
$$

This array contains two sets of four independent 0s, one corresponding to Ac, Bb, Ca, Dd and one corresponding to Ad, Bc, Ca, Db. These assignments both have a total unsuitability measure of 22, and are equally best possible assignments.

The procedure described here always yields n independent 0s after a finite number of repetitions. All that remains, therefore, is to use Hall's theorem to prove the result quoted and used in the description of the method.

THEOREM 3.5. (the König-Egerváry max–min theorem). *Let A be any $m \times n$ matrix. The maximum number of independent 0s which can be found in A is equal to the minimum number of lines (rows or columns) which together cover all the 0s of A.*

Proof. If the maximum number of independent 0s is k, it is obvious that no set of fewer than k lines can cover all the 0s.

$$\text{min. no. of lines covering 0s} \geqslant \text{max. no. of independent 0s.} \quad (3.4)$$

It will now be proved that if the minimum number of lines covering all the 0s is h, then there are h independent 0s. This will show that

$$\text{max. no. of independent 0s} \geqslant \text{min. no. of lines covering 0s,} \quad (3.5)$$

which, together with (3.4), will give the required result.

Suppose the h lines in a minimal cover consist of a rows and b columns. By changing the order of the rows and columns, these can be supposed to be the first a rows and the first b columns ($a + b = h$). Since all the 0s lie in these lines, A must be of the form

$$
A = \left(\begin{array}{c|c} C & D \\ \hline E & F \end{array} \right)_{m \times n}
$$

where F contains no 0s, C is $a \times b$, D is $a \times (n - b)$, E is $(m - a) \times b$ and F is $(m - a) \times (n - b)$.

It will now be shown that D contains a independent 0s. Then E will similarly contain b independent 0s. Neither of these sets of 0s can interfere with the other, and so there will be $a + b = h$ independent 0s in A as required.

If D does not have a independent 0s then, by Hall's criterion, there must be k rows of D, for some k, all the 0s in which lie in $t < k$ columns. But these 0s could have been covered by these t columns instead of the k rows. This gives a cover of the 0s which is smaller than the minimal cover, on replacing the k rows by the $t < k$ columns and leaving the other lines of the cover unchanged. This impossible situation implies that the assumption about D must have been wrong. Thus D does have a independent 0s, and the proof is complete.

This max–min theorem is in fact equivalent to Hall's in the sense that not only can it be deduced from Hall's theorem, but Hall's theorem can be deduced from it. (See Exercises 3.3, question 3.) One of the most fascinating regions of combinatorial mathematics is the study of theorems which are equivalent to Hall's theorem. Because many such theorems exist, taken together they must be considered to be the basic theorem in the whole of the subject.

Exercises 3.3

1. Five men live in towns A, B, C, D, E and they have to visit towns a, b, c, d, e, one man going to each town. The distances are given in the following table. Which towns should be visited by which men if the total amount of travelling is to be minimized?

	A	B	C	D	E
a	50	41	50	85	42
b	110	122	54	147	57
c	127	78	151	99	172
d	109	67	144	73	160
e	52	102	117	89	49

2. Write down a 5 x 6 matrix, and verify the max–min theorem for it.

3. Deduce Hall's theorem from the max–min theorem. (Hint: define the matrix $A = (a_{ij})$ by putting $a_{ij} = 0$ if the ith element is in the jth set, and $a_{ij} = 1$ otherwise.)

4. Four girls and four boys apply to a computer matching service, and the computer evaluates the suitability of each girl for each boy as a percentage. The respective suitabilities are shown in the accompanying table. Write down the unsuitability table (unsuitability = 100 − suitability) and hence find the best possible matching.

	B_1	B_2	B_3	B_4
G_1	40	65	70	45
G_2	70	90	45	70
G_3	60	40	85	65
G_4	75	85	60	60

5. In a college mathematics department, one of six lecturers has to be assigned to each of five classes. An examination of the suitability of each lecturer for each class is carried out, and the unsuitability table is produced. What is the optimal assignment if no lecturer is to be assigned to more than one class?

	L_1	L_2	L_3	L_4	L_5	L_6
C_1	70	20	50	20	45	30
C_2	55	25	40	30	40	15
C_3	50	20	45	10	45	25
C_4	35	30	35	40	30	20
C_5	45	15	40	25	35	40

(Hint: add a further class C_6 for which each lecturer is equally suited.)

3.4. Gale's optimal assignment problem

Yet another problem of the assignment type arises when the jobs to be filled are listed in order of importance, and it is desired to assign men to jobs in the best possible way. As for Hall's theorem, each man is suited for only certain of the jobs.

It is not at all obvious that a reasonable solution to this problem can be found. For example, it may be possible to appoint men to jobs 1, 2, and 4 or to jobs 2, 3, 5, and 6, where the jobs are in decreasing order of importance, and it is open to argument which of these is to be preferred. The latter assignment fills more jobs, but the former fills more important ones. Which is better? Fortunately a decision of this type need never be taken in practice. Call a set of jobs *assignable* if different men can be assigned to each, and suppose that the jobs are always listed in decreasing order of importance. It will be shown that there always is an assignable set $\{a_1, \ldots, a_n\}$ which is *optimal* in the sense that if $\{b_1, \ldots, b_m\}$ is any other assignable set then $m \leqslant n$ and $b_i \geqslant a_i$ for each $i \leqslant m$. Here $b_i > a_i$ means that job b_i is less important than a_i.

To illustrate this idea, suppose that the following is a list of 5 jobs $J_1, \ldots, J_5$, along with the men suited for each.

J_1: suitable men are A, B
J_2 B, C
J_3 B
J_4 A, C
J_5 B, C, D

The optimal assignable set of jobs is $\{J_1, J_2, J_3, J_5\}$. This set is assignable since the jobs in it can be filled by A, C, B, D respectively. Other assignable sets are, for example, $\{J_2, J_3, J_4\}$ and $\{J_1, J_4, J_5\}$.

How can an optimal assignable set of jobs always be shown to exist? It is necessary first to show that in assignment problems the assignable sets possess what is known as the *exchange property*. This property is simply the following. If $\{a_1, \ldots, a_r\}$ and $\{b_1, \ldots, b_{r+1}\}$ are two assignable sets, then the first set can be extended to an assignable set of $(r + 1)$ jobs by adding to it one of the members of the second set not yet in it. In the above example, $\{J_1, J_2\}$ and $\{J_2, J_3, J_4\}$ are two assignable sets, and the smaller can be extended to $\{J_1, J_2, J_3\}$ which is also assignable. The sets have *exchanged* J_3. This exchange property is characteristic of systems known as *matroids*, which are becoming one of the most important structures in combinatorial theory. No further mention will be made of matroids in this book, but the reader is recommended to have a look at the introduction given in [7]. Also, Gale's proof of the existence of an optimal assignable set of jobs [12], which differs in some of its aspects from the present treatment, is couched in terms of matroids, and is an excellent means of being introduced to them.

Once the exchange property has been established, it is very easy to prove that an optimal assignable set exists; indeed it is easy to see how to obtain it. Simply follow the procedure described in the proof of Hall's theorem, dealing with the jobs in decreasing order of importance. At the rth stage, first see if it is possible to assign the rth job to a man who does not yet have a job. If so, assign him to it. If not, try to alter the previous assignments as described in the algorithm. If the method fails, abandon the job (it will remain for ever unfillled) and proceed to the $(r + 1)$th job.

The set of jobs $a_1, \ldots, a_n$ obtained in this way is optimal in the sense described above. For suppose that $\{b_1, \ldots, b_m\}$ is any other assignable set. First, $m \leqslant n$, for if $m > n$ then, by the exchange property, one of the bs could be added to the set of as to give another assignable set. But this contradicts the fact that each b not in the set of as was rejected in the construction. Secondly, $b_i \geqslant a_i$ for each $i \leqslant m$. For suppose that $b_j < a_j$ and apply the exchange property to the two sets $\{a_1, \ldots, a_{j-1}\}$ and

$\{b_1, \ldots, b_j\}$. There must be a job b_k with $k \leqslant j$ such that $\{a_1, \ldots, a_{j-1},$ $b_k\}$ is assignable. But b_k, being at least as important as b_j, is more important than a_j, and so must have been rejected in the construction. Thus $\{a_1, \ldots, a_{j-1}, b_k\}$ cannot possibly be an assignable set, giving a contradiction. Thus the constructed set has been shown to be optimal.

It remains only to prove the exchange property.

Exchange property

Let $\{a_1, \ldots, a_n\}$ and $\{b_1, \ldots, b_{n+1}\}$ be two assignable sets, not necessarily disjoint. Then, for some $i \leqslant n + 1$, $\{a_1, \ldots, a_n, b_i\}$ is also an assignable set.

Proof. If $n = 1$, the reader can see that the property is trivial. If a_1 is in fact one of b_1, b_2 then the set $\{b_1, b_2\}$ is an extending set as required. If a_1 is different from both b_1 and b_2, one of the jobs b_1, b_2 must be filled by a man who is not doing job a_1, so that his job can be added to a_1. The proof now proceeds by induction, showing that if the property is true whenever $n = k - 1$ then it is true whenever $n = k$. So suppose that $\{a_1, \ldots, a_k\}$ and $\{b_1, \ldots, b_{k+1}\}$ are assignable sets of jobs, and that men have been assigned to the jobs in each case. One of the bs, say b_h, must be filled by a man who does not have any of the jobs a_i, there being more bs than as. If the job b_h is not one of the as, then it can be added to the set of as and the proof is complete. If b_h is one of the as, say a_t, then reassign job a_t to the man who is doing job b_h, and remove this job from both lists. The sets of jobs now consist of $(k - 1)$ and k jobs respectively, and by the induction hypothesis one of the bs can be added to the set of as. Finally add a_t back into the set to obtain as assignable set as required.

Worked example. We now illustrate the above theory by obtaining the optimal assignment in the example mentioned in the text. There are five jobs $J_1, \ldots, J_5$ as follows:

$$
\begin{aligned}
J_1 &: \text{suitable men are A, B} \\
J_2 &: \qquad\qquad\qquad \text{B, C} \\
J_3 &: \qquad\qquad\qquad \text{B} \\
J_4 &: \qquad\qquad\qquad \text{A, C} \\
J_5 &: \qquad\qquad\qquad \text{B, C, D.}
\end{aligned}
$$

Start by assigning A to job J_1 and then B to job J_2. Job J_3 cannot be filled by a remaining man, so proceed as in the algorithm for Hall's theorem. Diagrammatically we have

$$
\begin{array}{cc}
\text{B} & \text{C} \\
| & \\
J_3 & J_2
\end{array}
$$

so the assignment should be changed to J_3 filled by B, J_2 by C, with J_1 filled by A as before. Now consider J_4. Possible men for it are A and ᐧC, neither of whom is free. Proceed as in Hall's theorem to try to obtain an alternative assignment:

$$
\begin{array}{ccc}
A & B & C \\
| & | & | \\
J_4 & J_1 & J_3 & J_2.
\end{array}
$$

The method breaks down at this point, so the possibility of finding a man for J_4 is abandoned. Finally, J_5 can be filled immediately by D. The optimal solution is therefore to fill jobs J_1, J_2, J_3, J_5.

Exercises 3.4

1. Find the optimal assignable set in Example 3.4 in the text of § 3.2 if the jobs are in decreasing order of importance.

2. Repeat the previous question with the jobs in increasing order of importance.

3. A collection $\mathscr{C}$ of subsets of a set S is a *matroid* if
   (a) any subset of a set in $\mathscr{C}$ is also in $\mathscr{C}$,
   (b) if $\{a_1, \ldots, a_k\}$ and $\{b_1, \ldots, b_{k+1}\}$ are both in $\mathscr{C}$, then so is $\{a_1, \ldots, a_k, b_i\}$ for some $i \leqslant k + 1$.
   If $S = \{1, \ldots, 7\}$ and $\mathscr{C}$ consists of all subsets of S with $\leqslant 2$ elements and all the sets with 3 elements excluding $\{1, 2, 4\}$, $\{2, 3, 5\}$, $\{3, 4, 6\}$, $\{4, 5, 7\}$, $\{5, 6, 1\}$, $\{6, 7, 2\}$, $\{7, 1, 3\}$, show that $\mathscr{C}$ is a matroid. The reader will find it instructive to return to this example after studying Example 6.1.

4. A set B in a matroid M is said to be *maximal* if no other set in M contains B as a subset. Use the exchange property to prove that all maximal sets in M have the same number of elements.

5. Given a graph G, let M be the set of all sets of edges of G which do not contain a cycle (see page 38). Show that M is a matroid.

3.5. Further reading on Chapter 3

For a full study of Hall's marriage problem and its equivalent formulations, see the survey article of Mirsky and Perfect [19] or the book by Mirsky [4].

Theorem 3.2 gives a sufficient condition for a graph to possess a perfect matching, but this condition is not necessary. Tutte [23] in 1947 published a necessary and sufficient condition, and a simple proof of this result, due to the present author and using only Hall's theorem, is given in [8].

Fisher's book [1] gives an elementary introduction to the use of Latin squares in the design of experiments.

4 Recurrence

4.1. Some miscellaneous problems

SOME combinatorial problems reduce to examining a sequence $\{a_n\}$ of numbers $a_1, a_2, a_3, \ldots$ in the hope of obtaining a formula for the nth member a_n of the sequence. Often a_n is expressed in terms of previous members of the sequence, i.e. a *recurrence relation* is given, and also the first few values are given, for example a_1 and a_2. The problem is then to deduce a formula for a_n.

A few such problems are now exhibited.

Example 4.1. The *Fibonacci sequence*, mentioned in Chapter 1, is defined by

$$a_1 = 1, \quad a_2 = 2, \quad a_n = a_{n-1} + a_{n-2} \quad (n \geqslant 3),$$

and the problem is to find a formula for a_n. This sequence was investigated in the 13th century by Leonardo Fibonacci of Pisa, in connection with the growth of the rabbit population. (See Exercises 4.1.)

Example 4.2. Some combinatorial problems in chemistry reduce to counting the number of graphs of a certain type. A *tree* is defined to be a connected graph with no cycles, i.e. a connected graph in which it is impossible to start at a vertex, move along different edges and arrive back at the starting place. Examples of trees are shown in Fig. 4.1,

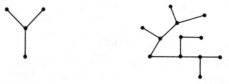

FIG. 4.1

whereas the graph in Fig. 4.2 is not a tree, one cycle being *abcd*. Trees can be used to represent the structure of chemical compounds, and it was in this way that Cayley was led to his studies of graph theory in the 1870s.

As an example of the type of problem involved, consider the problem of counting simple rooted trees. A *simple* tree is defined to be a tree in

FIG. 4.2

which each vertex is of degree $\leqslant 3$. (Recall that the degree of a vertex is the number of edges emanating from it.) One way of looking at a simple tree is to consider it as a road system in which one has a choice of at most two roads at each roadend. The simple trees to be considered are those *rooted* at a certain vertex P (see Fig. 4.3). P can be considered as the

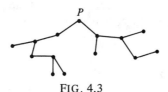

FIG. 4.3

starting point of the tree's growth, and in accordance with the requirement that at most two edges are available on reaching any vertex, it will be assumed that there are at most two edges emanating from P. An example of such a rooted simple tree is shown. The problem is to evaluate u_n, the number of different rooted simple trees with n vertices.

A difficulty, fundamental to most combinatorial problems, immediately arises. When are two trees to be considered different? For example, are the two trees in Fig. 4.4 the same or different? After all, in any practical realization, (a) can be picked up and turned over to give (b).

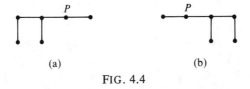

(a)   (b)

FIG. 4.4

Since it is a good idea to start with as simple a problem as possible, it will be considered here that (a) and (b) are distinct.† Then

† If (a) and (b) are considered the same, the resulting counting problem is much more difficult, and the ideas of Polya's theorem are required. See, for example, Harary, *Graph theory* [3].

$u_1 = 1$

$u_2 = 1$

$u_3 = 2$

$u_4 = 4$

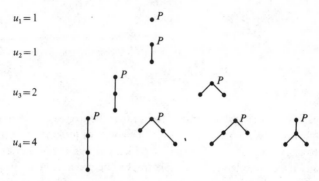

and so on. In evaluating u_n, $n \geqslant 2$, two types of tree have to be considered. Let s_n denote the number with only one edge from the root P, and let d_n denote the number with two edges from P. Clearly

$$u_n = s_n + d_n. \quad (n \geqslant 2) \tag{4.1}$$

Now consider s_{n+1}. A tree contributing to s_{n+1} is of the form

where, inside the circle, there can be any simple tree with n vertices rooted at Q. There are u_n such trees. Thus

$$s_{n+1} = u_n. \tag{4.2}$$

Next consider d_{n+1}. A tree contributing to d_{n+1} is of the form

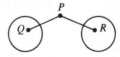

where there is a rooted simple tree at Q with, say, r vertices, and a rooted simple tree at R with s vertices, $r + s = n, r \geqslant 1, s \geqslant 1$. For each such pair of values of r and s there are u_r choices of what happens at Q and u_s choices at R; $u_r u_s$ choices altogether. Thus

$$d_{n+1} = u_1 u_{n-1} + u_2 u_{n-2} + \cdots + u_{n-1} u_1$$

$$= \sum_{\substack{r+s=n \\ r \geqslant 1, s \geqslant 1}} u_r u_s. \tag{4.3}$$

The relations (4.1) to (4.3) then yield

$$u_{n+1} = s_{n+1} + d_{n+1}$$

$$= u_n + \sum_{\substack{r+s=n \\ r \geqslant 1, \, s \geqslant 1}} u_r u_s,$$

i.e.

$$u_n = u_{n-1} + (u_1 u_{n-2} + u_2 u_{n-3} + \cdots + u_{n-2} u_1). \tag{4.4}$$

How is a formula for u_n obtained from this recurrence relation?

Example 4.3. *The problem of derangements.* Suppose that n jobs have been assigned to n people. In how many ways can they be reassigned the following day so that no person is given the same job as before?

In general, a *derangement* of the numbers $1, 2, \ldots, n$ is a rearrangement or permutation of them such that no number appears in its original position. For example, 23514 is a derangement of 12345, but 23541 is not. Let a_n denote the required number; then a_n is simply the number of derangements of $1, \ldots, n$, for it can be supposed that the jobs are so labelled that the ith person got the ith job on the first day.

Clearly $a_1 = 0$ (why?), $a_2 = 1$, $a_3 = 2$. To see that $a_3 = 2$, note that the only derangements of 123 are 231 and 312.

Suppose now that $n > 2$, and consider two possibilities. The first possibility is that in a derangement of $1, \ldots, n$ the number n changes places with some other number r. There are $(n-1)$ choices for r, and for each such choice the remaining $(n-2)$ numbers must undergo a derangement. The number of ways this can happen is $(n-1)a_{n-2}$, and this therefore gives the number of derangements of $1, \ldots, n$ in which n changes places with another number. The second possibility to consider is when some number r moves to the nth place, but n does not move to the rth place. In this case, ignore r which has now been placed, and relabel n by r. This gives $(n-1)$ numbers $1, \ldots, (n-1)$ to arrange, and the condition is again simply that no i is to placed in the ith place. There are a_{n-1} such derangements for each of the $n-1$ choices of r, and so $(n-1)a_{n-1}$ derangements of this type. Thus

$$a_n = (n-1)a_{n-1} + (n-1)a_{n-2}. \tag{4.5}$$

and the problem is how to solve this recurrence relation subject to the boundary conditions $a_1 = 0$, $a_2 = 1$.

Exercises 4.1

1. Use (4.4) to find u_5, and check your answer by drawing all possible rooted simple trees with 5 vertices.

2. Use (4.5) to find a_4, and check your answer by writing down all the possible derangements of 1234.

3. Suppose that any newborn pair of rabbits will produce their first pair of offspring after two months, and thereafter will produce one pair per month. Starting with one newborn pair, the growth of population is as follows, where A denotes a newborn pair, B a month-old pair, and C a fully-adult pair:

$$
\begin{array}{ll}
\text{after 1 month} & B \\
\qquad\ \ 2 \text{ months} & C\,A \\
\qquad\ \ 3 & C\,B\,A \\
\qquad\ \ 4 & C\,C\,B\,A\,A
\end{array}
$$

Prove that a_n, the number of pairs of rabbits in the population after n months, satisfies $a_1 = 1, a_2 = 2, a_n = a_{n-1} + a_{n-2}\ (n \geqslant 3)$. It is to be assumed that no deaths occur!

4.2. Fibonacci-type relations

A method of solving recurrence relations of the form

$$a_n = A a_{n-1} + B a_{n-2} \quad (n \geqslant 3) \tag{4.6}$$

is now given, where A and B are constants, i.e. do not depend on n. As is shown, the method is essentially just that of solving the associated quadratic equation

$$x^2 = Ax + B.$$

THEOREM 4.1. *Suppose that a_1 and a_2 are given and that (4.6) holds. Then*
(1) *if the roots α, β of the equation $x^2 = Ax + B$ are distinct, then*

$$a_n = K_1 \alpha^n + K_2 \beta^n,$$

where the constants K_1, K_2 are determined uniquely by a_1 and a_2;
(2) *if $x^2 = Ax + B$ has repeated root α, then*

$$a_n = (K_1 + nK_2)\alpha^n.$$

Example 4.4. *The Fibonacci sequence.* Here $A = B = 1$, so consider the equation $x^2 = x + 1$. This has roots

$$\alpha = \tfrac{1}{2}(1 + \sqrt{5}), \quad \beta = \tfrac{1}{2}(1 - \sqrt{5})$$

so that

$$a_n = K_1 \left(\frac{1 + \sqrt{5}}{2}\right)^n + K_2 \left(\frac{1 - \sqrt{5}}{2}\right)^n$$

for some constants K_1, K_2. Since $a_1 = 1$ and $a_2 = 2$,

$$1 = K_1 \left(\frac{1 + \sqrt{5}}{2}\right) + K_2 \left(\frac{1 - \sqrt{5}}{2}\right)$$

and

$$2 = K_1 \left(\frac{3 + \sqrt{5}}{2}\right) + K_2 \left(\frac{3 - \sqrt{5}}{2}\right).$$

These give

$$K_1 = \frac{\sqrt{5} + 1}{2\sqrt{5}}, \quad K_2 = \frac{\sqrt{5} - 1}{2\sqrt{5}}$$

so that

$$a_n = \frac{1}{\sqrt{5}} \left(\frac{1 + \sqrt{5}}{2}\right)^{n+1} - \frac{1}{\sqrt{5}} \left(\frac{1 - \sqrt{5}}{2}\right)^{n+1}.$$

This may at first sight seem rather odd, since it is known that a_n must be an integer. However, all the $\sqrt{5}$ terms cancel out. The binomial theorem gives

$$a_n = \frac{1}{\sqrt{5}} \cdot \frac{1}{2^{n+1}} \left\{ \sum_{r=0}^{n+1} \binom{n+1}{r} 5^{r/2} - \sum_{r=0}^{n+1} \binom{n+1}{r} (-1)^r 5^{r/2} \right\}$$

$$= \frac{1}{\sqrt{5}} \cdot \frac{1}{2^n} \left\{ \binom{n+1}{1} 5^{1/2} + \binom{n+1}{3} 5^{3/2} + \binom{n+1}{5} 5^{5/2} + \cdots \right\}$$

$$= \frac{1}{2^n} \left\{ \binom{n+1}{1} + 5 \binom{n+1}{3} + 5^2 \binom{n+1}{5} + \cdots \right\},$$

which is an interesting result since it is not immediately obvious that the sum in the brackets must be divisible by 2^n.

Proof of Theorem 4.1. (1) Suppose the formula for a_n has been proved for all n up to but not including some number k. Then

$$a_{k-1} = K_1\alpha^{k-1} + K_2\beta^{k-1}, \quad a_{k-2} = K_1\alpha^{k-2} + K_2\beta^{k-2}.$$

Thus

$$a_k = Aa_{k-1} + Ba_{k-2}$$
$$= K_1(A\alpha^{k-1} + B\alpha^{k-2}) + K_2(A\beta^{k-1} + B\beta^{k-2})$$
$$= K_1\alpha^{k-2}(A\alpha + B) + K_2\beta^{k-2}(A\beta + B)$$
$$= K_1\alpha^{k-2}\alpha^2 + K_2\beta^{k-2}\beta^2$$
$$= K_1\alpha^k + K_2\beta^k.$$

The formula therefore also holds for a_k.
(2) The proof is similar, and is left to the reader.

Exercises 4.2

1. If $a_n = 4(a_{n-1} - a_{n-2})$ for each $n \geqslant 3$, and if $a_1 = 0$, $a_2 = 4$, find a_n.

2. If $a_n = 5a_{n-1} - 6a_{n-2}$ for each $n \geqslant 3$, and if $a_1 = a_2 = 1$, find a_n.

3. If a_n denotes the nth Fibonacci number, prove that

$$a_{n+2} = a_n + a_{n-1} + \cdots + a_1 + 2.$$

4. Let $b_n = \binom{n}{0} + \binom{n-1}{1} + \binom{n-2}{2} + \cdots$. Verify that $b_1 = 1$, $b_2 = 2$, and show that $b_n = b_{n-1} + b_{n-2}$ ($n \geqslant 3$). Thus b_n gives another formula for the Fibonacci numbers.

5. If a_n is the nth Fibonacci number, prove that

$$a_n^2 - a_{n-1}a_{n+1} = (-1)^n.$$

6. In working through a problem, a man is said to be at the nth stage if he is n steps from the solution. At any stage he has 5 choices. Two of these result in him going to the $(n-1)$th stage, and three of them are better in that they take him direct to the $(n-2)$th stage. Let a_n denote the number of ways he can reach the solution from the nth stage. If $a_1 = 2$, verify that $a_2 = 7$ and obtain a recurrence relation for a_n. Deduce that

$$a_n = \tfrac{1}{4}\{3^{n+1} + (-1)^n\}.$$

7. Complete the proof of Theorem 4.1.

8. If $a_n = A a_{n-1} + B a_{n-2}$, and if $S_n = a_1 + \cdots + a_n$, prove that $S_{n+3} = (A + 1)S_{n+2} + (B - A)S_{n+1} - B S_n$. Verify this in the case $A = B = 1$, where the a_n are the Fibonacci numbers.

9. The $n \times n$ determinant D_n is defined for $n \geq 1$ by

$$D_n = \begin{vmatrix} 1 + a^2 & a & 0 & 0 & \cdots & 0 \\ a & 1 + a^2 & a & 0 & \cdots & 0 \\ 0 & a & 1 + a^2 & a & \cdots & 0 \\ \vdots & & & & & \\ 0 & 0 & 0 & 0 & \cdots & 1 + a^2 \end{vmatrix}$$

Show that, if $n \geq 3$, $D_n = (1 + a^2)D_{n-1} - a^2 D_{n-2}$ and hence show that

$$D_n = \frac{1 - a^{2n+2}}{1 - a^2} \quad \text{if } a^2 \neq 1.$$

What if $a^2 = 1$?

10. Let a_n denote the number of n-digit sequences in which each digit is 0 or 1, no two consecutive 0s being allowed. Show that $a_1 = 2, a_2 = 3$ and that $a_n = a_{n-1} + a_{n-2} (n \geq 3)$. Hence find a_n.

11. Let b_n denote the number of n-digit sequences in which each digit is 0, 1, or − 1, if no two consecutive 1s or consecutive − 1s are allowed. Prove that $b_n = 2b_{n-1} + b_{n-2} (n \geq 3)$ and hence find b_n.

4.3. Using generating functions

The recurrence relation (4.4) obtained on counting rooted simple trees does not look too attractive; it looks too difficult to deal with. It sometimes happens that such relations are better dealt with by means of generating functions. As was explained in the opening chapter, the generating function for a given sequence $a_0, a_1, a_2, \ldots, a_n, \ldots$ is defined to be

$$f(x) = a_0 + a_1 x + a_2 x^2 + \cdots + a_n x^n + \cdots,$$

where the coefficient of x^n in $f(x)$ is precisely the term a_n of the sequence.

Let $u(x), s(x)$, and $d(x)$ be the generating functions for Example 4.2, where

$$u(x) = u_1 x + u_2 x^2 + u_3 x^3 + \cdots$$
$$= x + x^2 + 2x^3 + \cdots,$$

$$s(x) = s_1 x + s_2 x^2 + s_3 x^3 + \cdots$$
$$= x^2 + x^3 + 2x^4 + \cdots,$$
$$d(x) = d_1 x + d_2 x^2 + d_3 x^3 + \cdots$$
$$= x^3 + 2x^4 + \cdots,$$

on noting that $s_1 = d_1 = d_2 = 0$ (why?). From (4.1),

$$u(x) = x + s(x) + d(x). \tag{4.7}$$

Also, since $s_{n+1} = u_n$,

$$s(x) = xu(x), \tag{4.8}$$

as can be checked by comparing the coefficients of any power of x on each side of the equation. Finally, from (4.3), it follows that

$$d(x) = x\{u(x)\}^2. \tag{4.9}$$

Eqns (4.7)–(4.9) together give

$$u(x) = x + xu(x) + x\{u(x)\}^2,$$

i.e.

$$x\{u(x)\}^2 + (x - 1)u(x) + x = 0, \tag{4.10}$$

which is a quadratic equation for $u(x)$. The usual formula for solving such equations then gives

$$u(x) = \frac{1}{2x} [1 - x \pm \sqrt{\{(x - 1)^2 - 4x^2\}}]$$

$$= \frac{1}{2x} [1 - x \pm \sqrt{\{1 - (2x + 3x^2)\}}]. \tag{4.11}$$

Now, by the binomial theorem,

$$(1 - y)^{\frac{1}{2}} = 1 - \tfrac{1}{2}y - \frac{\tfrac{1}{2} \cdot \tfrac{1}{2}y^2}{2!} - \frac{\tfrac{1}{2} \cdot \tfrac{1}{2} \cdot \tfrac{3}{2}y^3}{3!} - \cdots$$

$$\cdots - \frac{1 . 3 . 5 \ldots (2n - 3)}{2^n n!} y^n - \cdots, \tag{4.12}$$

so that, on taking the minus sign in (4.11) to obtain positive coefficients u_n,

$$u(x) = \frac{1}{2x}\left[-x + \tfrac{1}{2}(2x + 3x^2) + \frac{1}{2^2 2!}(2x + 3x^2)^2 + \cdots\right]$$

$$= x + x^2 + 2x^3 + 4x^4 + \cdots. \tag{4.13}$$

Technically, the problem is now solved. To find u_n, all that need be done is to read off the coefficient of x^n in (4.13). If h_n denotes the numerical value of the coefficient of y^n in (4.12), so that

$$h_n = \frac{(2n - 2)!}{2^{2n-1}n! \, (n - 1)!}$$

then it is straightforward to verify that

$$u_{n-1} = \tfrac{1}{2}\left\{h_n 2^n + h_{n-1}2^{n-2} \cdot 3 \cdot \binom{n-1}{1} + h_{n-2}2^{n-4}3^2 \binom{n-2}{2} + \right.$$

$$\left. + \cdots + h_{n-r}2^{n-2r}3^r \binom{n-r}{r} + \cdots \right\}. \tag{4.14}$$

This formula for u_n has its unattractive side. It is not very compact, and a certain amount of effort is still required to evaluate u_n for any specific value of n, particularly when n is large. However, all that is involved is essentially the substitution of the particular value of n into (4.14), and this is all that is required of a formula. Some mathematicians would take the view that the problem of finding u_n was in fact solved well before this final formula was obtained—at the stage (4.13) of obtaining the generating function $u(x)$, since the values of all the coefficients u_n are implicit in $u(x)$. This book will take the view that an explicit formula is to be preferred to simply a generating function solution, and such a formula should be aimed at whenever possible.

Further examples on generating functions

Example 4.5. Suppose that, in the problem posed at the beginning of Chapter 1, there are 4 colours available (i.e. $n = 4$). How many colourings of the k golf balls are possible if there must be an odd number of objects coloured with the first colour?

Solution (1). As in the second approach to the original problem, the required number is the coefficient of x^k in

$$(x+x^3+x^5+\cdots)(1+x+x^2+x^3+\cdots)^3=x(1+x^2+x^4+\cdots)(1-x)^{-3},$$

which is just the coefficient of x^{k-1} in

$$(1+x^2+x^4+\cdots)(1+\tbinom{3}{1}x+\tbinom{4}{2}x^2+\tbinom{5}{3}x^3+\cdots).$$

This coefficient is

$$\binom{k+1}{k-1}+\binom{k-1}{k-3}+\binom{k-3}{k-5}+\cdots=\binom{k+1}{2}+\binom{k-1}{2}+\binom{k-3}{2}+\cdots.$$

For example, $k=6$ gives 34 possible colourings.

Solution (2). Alternatively, if exactly one ball is coloured with the first colour, there are $(k-1)$ balls left to be coloured with 3 colours. By the result (1.6), this can be done in $\binom{k-1+3-1}{k-1}=\binom{k+1}{k-1}$ ways. Similarly, if exactly 3 are coloured with the first colour, the remaining $(k-3)$ can be coloured in $\binom{k-3+3-1}{k-3}=\binom{k-1}{k-3}$ ways. Continuing in this way the same result is obtained as before.

Example 4.6. n-digit integer sequences are to be formed using only the integers 0, 1, 2, 3. For example, 0031 and 3202 are two 4-digit sequences.

(a) How many n-digit sequences are there?

(b) How many n-digit sequences have an odd number of 0s?

Solution. (a) The number of sequences is 4^n, since there are 4 choices for each of the n digits.

(b) This is not so easy. The difference between the problem posed here and Example 4.5 is that here it matters not only what digits appear, but also in what order they occur.

Any n-digit sequence will consist of d_0 0s, d_1 1s, d_2 2s and d_3 3s, where d_0 is odd and $d_0+d_1+d_2+d_3=n$. Any given set of ds satisfying these conditions will give rise to as many different sequences as there are ways of arranging the n numbers in a line. If the n numbers were all distinct, there would be $n!$ permutations. Thus, if the n digits are labelled so that digits of the same kind are distinguishable from one another, there are $n!$ permutations. However, two of these permutations will be the same when the labels are removed if and only if they differ only in the arrangement of the d_0 0s among themselves, the d_1 1s, the d_2 2s, and the d_3 3s. Thus each permutation of the unlabelled digits corresponds to $d_0! \, d_1!$ $d_2! \, d_3!$ permutations of the labelled digits. Thus the number of distinct sequences with d_0 0s, d_1 1s, d_2 2s, and d_3 3s is

$$\frac{n!}{d_0! \, d_1! \, d_2! \, d_3!}.$$

Hence the total number of sequences is equal to

$$\sum \frac{n!}{d_0! \, d_1! \, d_2! \, d_3!}, \tag{4.15}$$

where the sum is over all sets of numbers $d_0, \ldots, d_3$ such that d_0 is odd and $d_0 + d_1 + d_2 + d_3 = n$. On looking for a possible generating function, the factorials on the denominator lead one to try the exponential function $\exp(x)$ introduced in Example 2.10. So consider

$$\left(x + \frac{x^3}{3!} + \frac{x^5}{5!} + \cdots\right)\left(1 + x + \frac{x^2}{2!} + \frac{x^3}{3!} + \cdots\right)\left(1 + x + \frac{x^2}{2!} + \cdots\right)$$

$$\times \left(1 + x + \frac{x^2}{2!} + \cdots\right). \tag{4.16}$$

The coefficient of x^n is $\dfrac{1}{n!}$ times the number given by (4.15), as can be seen by considering the ways in which x^n can be obtained by selecting a term from each bracket and multiplying them together. But (4.16) is

$$\left(x + \frac{x^3}{3!} + \frac{x^5}{5!} + \cdots\right)\{\exp(x)\}^3 = \left(x + \frac{x^3}{3!} + \frac{x^5}{5!} + \cdots\right)\exp(3x)$$

(by Example 2.10)

$$= \tfrac{1}{2}(\exp(x) - \exp(-x))\exp(3x)$$

$$= \tfrac{1}{2}(\exp(4x) - \exp(2x)).$$

The coefficient of x^n in this is

$$\tfrac{1}{2}\left(\frac{4^n}{n!} - \frac{2^n}{n!}\right).$$

The number of sequences is $n!$ times this number, namely

$$\tfrac{1}{2}(4^n - 2^n).$$

For another method of solving this example, see Exercises 4.3, question 4. As far as the above solution is concerned, the reader should remember not so much the answer as the idea of making use of the properties of $\exp(x)$.

Example 4.7. *Partitions of an integer.* Ideas from number theory have the habit of appearing all over the place and when least expected. One such idea is that of a *partition* of an integer. By a partition of a positive integer n is meant the expression of n as a sum of positive integers. For example, 5 has seven partitions:

$$5 = 4 + 1 = 3 + 2 = 3 + 1 + 1 = 2 + 2 + 1$$
$$= 2 + 1 + 1 + 1 = 1 + 1 + 1 + 1 + 1.$$

Note that 5 itself is a partition of 5. Let $p(n)$ denote the number of partitions of n, so that $p(5) = 7$, and let $f(x)$ be the generating function,

$$f(x) = p(1)x + p(2)x^2 + \cdots + p(n)x^n + \cdots.$$

Consider the expression

$$(1 - x)^{-1}(1 - x^2)^{-1}(1 - x^3)^{-1} \cdots$$
$$= (1 + x + x^2 + \cdots)(1 + x^2 + x^4 + \cdots)(1 + x^3 + x^6 + \cdots) \dots.$$

What is the coefficient of x^n in this expression? Note that $n$xs can be obtained by selecting a power of x from the first bracket, another from the second, and so on, and multiplying them together. Thus if x^{ia_i} is chosen from the ith bracket, x^n will be obtained if n is the sum of a_1 1s, a_2 2s, and so on. Thus x^n will be obtained as many times as n has different partitions, so that the coefficient of x^n must be $p(n)$. This proves that the generating function is

$$f(x) = (1 - x)^{-1}(1 - x^2)^{-1}(1 - x^3)^{-1} \dots.$$

Although such a generating function does not yield a formula for $p(n)$ easily, it turns out to be useful enough to yield properties of partitions. For an example of this, see the exercises below. Further, the final section of this chapter will give an illustration of how partitions appear naturally in a physical combinatorial problem.

Exercises 4.3

1. Let $f(x)$ denote the generating function of the Fibonacci numbers. Show that the recurrence relation gives

$$f(x) = x + 2x^2 + x(f(x) - x) + x^2 f(x),$$

so that

$$(1 - x - x^2)f(x) = x + x^2.$$

Deduce that

$$f(x) = (x + x^2)\{1 - (x + x^2)\}^{-1}$$
$$= (x + x^2)\{1 + x(1 + x) + x^2(1 + x)^2 + \cdots\}.$$

Read off the coefficient of x^n in this expression, and check that your answer agrees with Exercises 4.2, question 4.

2. Suppose that n objects lie in a straight line. Two adjacent objects are chosen and bracketed together, and thereafter are considered as just one object. This results in $(n - 1)$ objects in a line. Two of these $(n - 1)$ objects which are adjacent are then bracketed together and thereafter considered as just one object. This process is continued until only one object is left. Let a_n denote the number of ways the process can be carried out, starting with n objects, so that $a_1 = 1$, $a_2 = 1, a_3 = 2$. By observing that in the last bracketing there are grouped together r original objects and $(n - r)$ original objects, for some r, show that

$$a_n = a_1 a_{n-1} + a_2 a_{n-2} + \cdots + a_{n-1} a_1 \quad (n \geqslant 3).$$

Deduce that the generating function $f(x)$ satisfies

$$\{f(x)\}^2 - f(x) + x = 0,$$

and hence show that

$$a_n = \frac{(2n - 2)!}{n! (n - 1)!}.$$

3. Show that the number of ways of placing m similar parcels into n different boxes so that no box is empty is $\binom{m - 1}{n - 1}$.

4. Solve Example 4.6(b) as follows. Let a_n be the required number of n-digit sequences. By considering whether or not a given sequence begins with a 0, show that

$$a_{n+1} = 3a_n + (4^n - a_n), \text{ i.e. } a_{n+1} = 2a_n + 4^n.$$

Put $f(x) = \sum\limits_{n=1}^{\infty} a_n x^n$ and show that

$$f(x) = \frac{x}{(1 - 2x)(1 - 4x)},$$

whence $a_n = \frac{1}{2}(4^n - 2^n)$.

5. Let b_n denote the number of ways in which the sum n can be obtained on rolling a die any number of times. Show that the generating function for the b_i is

$$(1 - x - x^2 - x^3 - x^4 - x^5 - x^6)^{-1}.$$

6. (Harary and Read (1970). *Proc. Edinburgh math. Soc.*). Certain organic chemical compounds built up from benzene rings can be represented by hexagons joined together:

Benzene   Naphthalene   Anthracene   Phenanthracine

This raises the question: how many ways are there of combining together n hexagons? Simplify the problem as follows. First do not allow three hexagons to have a vertex in common. This means, for example, that a third hexagon cannot be nestled under two of the anthracene hexagons. Secondly, suppose that the configurations are all growing from a fixed spot, so that there is one fixed base hexagon. Onto this hexagon can be fitted either one hexagon (on any of the sides a, b, c) or two hexagons (one on each of a and c).

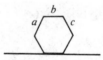

Let h_n denote the total number of possible patterns with n hexagons. Let s_n, d_n denote respectively the number with one, two hexagons joined to the base hexagon. Show that
(a) $s_n + d_n = h_n$ $(n \geqslant 2)$,
(b) $s_{n+1} = 3h_n$,
(c) $d_{n+1} = h_1 h_{n-1} + h_2 h_{n-2} + \cdots + h_{n-1} h_1$.
If $h(x)$, $s(x)$, $d(x)$ are the respective generating functions, deduce that
(d) $h(x) = s(x) + d(x) + x$,
(e) $s(x) = 3xh(x)$,
(f) $d(x) = x\{h(x)\}^2$

and that

$$x\{h(x)\}^2 + (3x - 1)h(x) + x = 0.$$

7. Let $q(n)$ denote the number of partitions of n into distinct parts. Thus $q(5) = 3$, since 5 can be written as 5 or $(4 + 1)$ or $(3 + 2)$. Show that the generating function $Q(x)$ is

$$(1 + x)(1 + x^2)(1 + x^3)(1 + x^4) \cdots.$$

8. Let $r(n)$ denote the number of partitions of n into odd parts. Thus $r(5) = 3$ since $5 = 3 + 1 + 1 = 1 + 1 + 1 + 1 + 1$. Show that the generating function $R(x)$ is

$$(1 - x)^{-1}(1 - x^3)^{-1}(1 - x^5)^{-1} \cdots.$$

9. Prove the surprising result that, in the notation of the previous exercises, $q(n) = r(n)$ for every value of n. Do this, without finding what $q(n)$ or $r(n)$ is, by showing that the generating functions are the same. (Hint: in $Q(x)$ write $1 + x^r$ as $\dfrac{1 - x^{2r}}{1 - x^r}$ and see what happens.)

10. Let $f(x)$ be the generating function for the sequence $a_1, a_2, \ldots$. Find the sequence whose generating function is $(1 - x)f(x)$. The answer should explain why $(1 - x)$ is called the *difference operator*.

11. The sequence $\{a_n\}$ is defined by $a_0 = e, a_1 = 2e$,

$$na_n = 2(a_{n-1} + a_{n-2}), \quad (n \geqslant 2).$$

Show that the generating function f satisfies the equation $f'(x) = 2(1 + x)f(x)$ and deduce that $f(x) = \exp\{(1 + x)^2\}$. Hence show that

$$a_{2n} = \sum_{r=0}^{\infty} \frac{1}{(n + r)!} \binom{2n + 2r}{2n},$$

$$a_{2n+1} = \sum_{r=0}^{\infty} \frac{1}{(n + r + 1)!} \cdot \binom{2n + 2r + 2}{2n + 1}.$$

12. Let a_n denote the number of ways in which n letters can be selected from the alphabet $\{0, 1, 2\}$ with unlimited repetitions except that the letter 0 must be selected an even number of times. Find a_n. How many n-letter sequences can be formed from this alphabet containing an even number of 0s?

4.4. Miscellaneous methods

The first recurrence relation mentioned in this book was

$$f(n, k) = f(n - 1, k) + f(n, k - 1) \qquad (4.17)$$

subject to the boundary conditions

$$f(1, k) = 1, \quad f(n, 1) = n. \qquad (4.18)$$

This has certain similarities to the recurrence relation for binomial coefficients,

$$\binom{n}{k} = \binom{n - 1}{k} + \binom{n - 1}{k - 1} \qquad (4.19)$$

subject to the boundary conditions

$$\binom{n}{0} = 1, \quad \binom{n}{n} = 1.$$

No general method will be given for dealing with relations such as (4.17), but it will be shown how (4.17) can be solved by exploiting its similarity to the known relation (4.19). In (4.17) the k terms behave as in (4.19), but the n terms do not. Diagrammatically, the pattern is

$$
\begin{array}{ccc}
n & n - 1 & n \\
k & k & k - 1
\end{array}
$$

compared with the binomial coefficient pattern of

$$
\begin{array}{ccc}
n & n - 1 & n - 1 \\
k & k & k - 1
\end{array}
$$

How can (4.17) be fitted into the required shape? Suppose the new function g is defined by

$$f(n, k) = g(n + k, k).$$

Then (4.17) becomes

$$g(n + k, k) = g(n + k - 1, k) + g(n + k - 1, k - 1) \qquad (4.20)$$

subject to the boundary conditions

$$g(1 + k, k) = 1, \quad g(n + 1, 1) = n.$$

Now (4.20) is familiar. For, with $m = n + k$, it is just

$$g(m, k) = g(m - 1, k) + g(m - 1, k - 1),$$

the recurrence relation for binomial coefficients. The boundary conditions, however, are not quite right. They would be, though, if the first variable were reduced by 1. So instead of putting $m = n + k$, try putting $u = n + k - 1$, and defining the function h by

$$f(n, k) = h(n + k - 1, k) = h(u, k).$$

(4.17) now becomes

$$h(u, k) = h(u - 1, k) + h(u - 1, k - 1)$$

subject to

$$h(k, k) = 1, \quad h(n, 1) = n.$$

It therefore follows that $h(u, k)$ must be $\binom{u}{k}$ so that, finally,

$$f(n, k) = \binom{n + k - 1}{k}.$$

Derangements

It has already been shown that if a_n denotes the number of derangements of n objects, then a_n satisfies the recurrence relation

$$a_n = (n - 1)a_{n-1} + (n - 1)a_{n-2}. \tag{4.21}$$

This is not one of the relations covered by Theorem 4.1 since the coefficients of a_{n-1} and a_{n-2} are not constants but depend on n. How can (4.21) be solved? One idea is to make a suitable substitution which will transform (4.21) into something more tractable. Define a new sequence $\{b_n\}$ by writing

$$a_n = n! \, b_n.$$

(4.21) then becomes

$$nb_n = (n - 1)b_{n-1} + b_{n-2} \quad (n \geqslant 3)$$

and the boundary conditions $a_1 = 0, a_2 = 1$, become

$$b_1 = 0, \quad b_2 = \tfrac{1}{2}.$$

This new relation does not look much better than the original until it is observed that it can be written as

$$n(b_n - b_{n-1}) = -(b_{n-1} - b_{n-2})$$

which, on putting $c_n = b_n - b_{n-1}$, becomes

$$c_n = -\frac{1}{n} c_{n-1}, c_2 = \tfrac{1}{2}.$$

This is easily dealt with, for clearly

$$c_n = \frac{(-1)^n}{n!} \quad (n \geqslant 2),$$

so that

$$b_n = c_n + b_{n-1}$$
$$= c_n + (c_{n-1} + b_{n-2}) = c_n + c_{n-1} + b_{n-2}$$
$$\vdots$$
$$= c_n + c_{n-1} + \cdots + c_2 + b_1$$

$$= \sum_{r=2}^{n} \frac{(-1)^r}{r!}.$$

Thus

$$a_n = n! \sum_{r=2}^{n} \frac{(-1)^r}{r!} = n! \sum_{r=0}^{n} \frac{(-1)^r}{r!},$$

i.e.

$$a_n = n! \left\{ 1 - \frac{1}{1!} + \frac{1}{2!} - \frac{1}{3!} + \cdots + (-1)^n \frac{1}{n!} \right\}. \tag{4.22}$$

This formula for a_n will be derived by another method when the inclusion-exclusion principle is introduced in the next chapter.

Exercises 4.4

1. As in Exercises 1.1, question 5, let $g(n, k)$ denote the number of ways of placing k indistinguishable lions in n cages so that no cage contains

more than one lion and no two lions are put in consecutive cages. It has been shown that

$$g(n, k) = g(n - 2, k - 1) + g(n - 1, k).$$

Define a new function h by $g(n, k) = h(p, k)$ where $p = n - k + 1$. Show that the recurrence relation and boundary conditions reduce to those for binomial coefficients, and deduce that $g(n, k) = \binom{n - k + 1}{k}$.

2. One form of assignment problem is the following. Given a set of men, called $A_1, \ldots, A_n$, and a set of jobs $J_1, \ldots, J_m$, it is desired to give each man a job. Let $a(n, m)$ be the number of ways in which this can be done so that every job has at least one man working on it.
   (a) Verify that $a(n, m) = 0$ if $n < m$.
   (b) Show that $a(n, n) = n!$
   (c) Show that, if the requirement that every job be filled is omitted, the men can be assigned to jobs in m^n ways.
   (d) Note that any one of the assignments in (c) results in exactly r jobs being filled, for some $r \leqslant m$. Deduce that

   $$m^n = \sum_{r=1}^{m} \binom{m}{r} a(n, r).$$

(e) Recall from Exercises 2.3, question 13, that

$$\sum_{s=r}^{m} (-1)^{m-s} \binom{m}{s}\binom{s}{r} = \begin{cases} 1 & \text{if } r = m \\ 0 & \text{if } r < m. \end{cases}$$

Using this, deduce that

$$\sum_{s=1}^{m} (-1)^{m-s} \binom{m}{s} s^n = \sum_{s=1}^{m} (-1)^{m-s} \binom{m}{s} \sum_{r=1}^{s} \binom{s}{r} a(n, r)$$

$$= \sum_{r=1}^{m} a(n, r) \sum_{s=r}^{m} (-1)^{m-s} \binom{m}{s}\binom{s}{r}$$

$$= a(n, m).$$

Thus

$$a(n, m) = \sum_{s=1}^{m} (-1)^{m-s} \binom{m}{s} s^n.$$

Now use this formula to find $a(10, 5)$. $a(n, m)$ is the number of surjections from a set of n elements to a set of m elements. The numbers $S(n, m)$, defined by $S(n, m) = \dfrac{1}{m!} a(n, m)$ are known as the *Stirling numbers of the second kind.*

3. Sequences of 0s and 1s are to be formed, containing exactly m 0s and at most n 1s.

   (a) Show that the number of such sequences of length $m + n$ is $\dbinom{m + n}{m}$.

   (b) Deduce or show otherwise that the total number of possible sequences is $\dbinom{m + n + 1}{n}$. It may be helpful to recall the result of Exercises 2.3, question 12.

   (c) Suppose now that $m = n$, and that the further condition is added that at any stage in the sequence the number of 1s which have so far occurred does not exceed the number of 0s which have appeared thus far. Let $a(n)$ be the number of such sequences, and define $a(0)$ to be 1 for convenience. By considering where in such a sequence the number of 1s last equalled the number of 0s, show that, for $n \geqslant 2$,

$$a(n) = a(0)a(n - 1) + a(1)a(n - 2) + \cdots + a(n - 1)a(0).$$

Let $f(x) = \displaystyle\sum_{n=0}^{\infty} a(n)x^n$ and deduce that

$$x\{f(x)\}^2 = f(x) - 1.$$

Hence show that $a(n) = \dfrac{1}{n + 1}\dbinom{2n}{n}$.

4. Let $g(m, n)$ denote the number of permutations $a_1, a_2, \ldots, a_n$ of $1, \ldots, n$ such that $a_i \geqslant i$ for exactly m is. Verify that $g(1, n) = g(n, n) = 1$ for all values of n. It can be shown (see [24]) that

$$g(m, n) = mg(m, n - 1) + (n - m + 1)g(m - 1, n - 1).$$

Deduce that

$$g(m, n) = g(n + 1 - m, n)$$

and, by induction on n, that

$$g(m, n) = \sum_{r = 0}^{m - 1} (-1)^r \binom{n + 1}{r} (m - r)^n.$$

Check this formula by verifying that $g(2, 4) = 11$.

5. A sequence of polynomials $\{L_n(x)\}$ is defined by

$$L_0(x) = 1, \quad L_1(x) = 1 + x, \quad L_k(x) = L_{k-1}(x) + xL_{k-2}(x), \quad (k \geqslant 2).$$

Let $f(k, m)$ denote the coefficient of x^m in $L_k(x)$, and show that

$$f(k, m) = f(k - 1, m) + f(k - 2, m - 1).$$

Deduce that

$$f(k, m) = \binom{k - m + 1}{m},$$

so that, in particular,

$$L_{2n-1}(x) = \sum_r \binom{2n - r}{r} x^r.$$

4.5. Counting simple electrical networks†

This section will deal with a certain class of electrical networks between two terminals. At first sight, it might appear that such networks are graphs, but this is not the case since there may be more than one edge joining two vertices; for example, the network shown in the diagram is

not a graph. In the context of networks, edges will be called *elements*, and the problem to be considered is that of counting the number of different networks with n elements. In its entire generality this problem is quite impossible, and so from now on the restriction will be made that all networks to be considered will be *simple*† in the following sense.

(1) The network •——• is a simple network.

(2) A simple network with $n > 1$ elements is one formed by combining either in series or in parallel any two simple networks with r and s elements respectively, where $r \geqslant 1, s \geqslant 1, r + s = n$.

For example, there are four simple networks with 3 elements:

On the other hand, the following network with 5 elements is not simple.

† J. Riordan and C. E. Shannon; see reference [20].

Before proceeding any further, the following remark must be made. When parts of a network are connected in series or parallel, the order in which they are connected is electrically irrelevant. This is why the following networks were not included in the census of networks with 3 elements:

Let u_n denote the number of simple networks with n elements. Then $u_1 = 1, u_2 = 2, u_3 = 4$ and $u_4 = 10$ (verify). Now the networks contributing to u_n fall into two disjoint groups, namely those whose construction involved a series connection at the last stage, and those involving a parallel connection. For $n = 3$ there are two of each type. This is typical in the sense that if s_n, p_n denote the number with last connection in series, parallel respectively, then $s_n = p_n$ for all $n \geqslant 2$. For consider the construction of one which culminates in a series connection, and interchange 'series' and 'parallel' at each stage of the construction. A network of the other type is obtained. Since $s_n = p_n$, it follows that $u_n = 2s_n$ $(n \geqslant 2)$. The reader should now check that $s_4 = p_4 = 5, s_3 = p_3 = 2, s_2 = p_2 = 1$.

Consider all the simple networks with n elements. Any such network consists of r parts connected in series for some number r, $1 \leqslant r \leqslant n$. (The case $r = 1$ corresponds to the last connection in the construction being a parallel one.) Each of the r parts has as its last connection a connection in parallel. Suppose there are d_1 parts with 1 element, d_2 with 2 elements, and so on, where $n = d_1 + 2d_2 + 3d_3 + \cdots + nd_n$, and consider in particular those d_i parts with i elements. Each of these d_i parts can be any of p_i types. Since order in series does not matter, the number of choices for these d_i parts is, according to the problem studied in Chapter 1,

$$\binom{p_i + d_i - 1}{d_i}$$

(the different types corresponding to different colours). The total number of possible networks with d_i parts of i elements is therefore, on replacing p_i by s_i,

$$\binom{s_1 + d_1 - 1}{d_1}\binom{s_2 + d_2 - 1}{d_2} \cdots \binom{s_n + d_n - 1}{d_n}. \tag{4.23}$$

Thus, taking into account all possible choices of $d_1, \ldots, d_n$,

$$2s_n = u_n = \sum \binom{s_1 + d_1 - 1}{d_1} \cdots \binom{s_n + d_n - 1}{d_n}, \qquad (4.24)$$

where the sum is over all the sets of numbers d_i such that

$$1 . d_1 + 2 . d_2 + \cdots + n . d_n = n. \qquad (4.25)$$

In other words, the sum in (4.24) is over all possible partitions of n.

The relation (4.24) can be used to evaluate any s_n provided each of $s_1, \ldots, s_{n-1}$ is known. For example, s_5 will now be calculated. The partitions of 5 are

5	(corresponding to $d_5 = 1$, all other $d_i = 0$)
4 + 1	($d_1 = d_4 = 1$)
3 + 2	($d_2 = d_3 = 1$)
3 + 1 + 1	($d_1 = 2, d_3 = 1$)
2 + 2 + 1	($d_1 = 1, d_2 = 2$)
2 + 1 + 1 + 1	($d_1 = 3, d_2 = 1$)
1 + 1 + 1 + 1 + 1	($d_1 = 5$).

A useful fact to note is that

$$\binom{s_i + d_i - 1}{d_i} = \begin{cases} 1 & \text{if } d_i = 0 \\ s_i & \text{if } d_i = 1. \end{cases}$$

Thus

$$2s_5 = s_5 + s_1 s_4 + s_2 s_3 + s_3 \binom{s_1 + 1}{2} + s_1 \binom{s_2 + 1}{2}$$

$$+ s_2 \binom{s_1 + 2}{3} + \binom{s_1 + 4}{5},$$

i.e.

$$s_5 = 5 + 2 + 2 + 1 + 1 + 1 = 12.$$

Finally, (4.24) can be expressed in terms of generating functions as follows.

Consider

$$(1 - x)^{-s_1}(1 - x^2)^{-s_2}(1 - x^3)^{-s_3} \cdots.$$

$$= \left\{ 1 + \binom{s_1 + 1 - 1}{1} x + \binom{s_1 + 2 - 1}{2} x^2 + \cdots \right\} \left\{ 1 + \binom{s_2 + 1 - 1}{1} x^2 \right.$$

$$\left. + \binom{s_2 + 2 - 1}{2} x^4 + \cdots \right\} \cdots.$$

The coefficient of x^n is precisely the sum of expressions of the form (4.23) subject to (4.25). It follows that the generating function f for the numbers s_i is given by

$$f(x) = \tfrac{1}{2}(1 - x)^{-s_1}(1 - x^2)^{-s_2}(1 - x^3)^{-s_3} \cdots,$$

where $s_0 = 1$ by convention. This generating function differs from those met so far in that the numbers s_n generated as coefficients are themselves involved in the generating expression. However, as shown in the evalution of s_5 above, this does not prevent the generating function being usable in calculating the coefficients. The reader should now show that $s_6 = 33$ and $s_7 = 90$.

5 The inclusion-exclusion principle

5.1. The principle

THERE are in mathematics a handful of principles which look so simple as to be valueless, but yet in practice are of the utmost importance and power. One such principle is the box principle which asserts that if $(n + 1)$ lions are put into n cages, then at least one cage must contain more than one lion. A course in number theory will show how powerful this simple principle is. The principle which is the subject of the present chapter is not much more difficult to understand. In its simplest form it is concerned with the number of elements in the union of two sets A and B (see Fig. 5.1). Let $|A|$ denote the number of elements in the set A. In

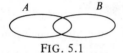

FIG. 5.1

evaluating $|A \cup B|$, consider first the possible answer $|A| + |B|$. This will in general be the wrong answer since those elements which are in both A and B are included twice, and therefore must be removed once. Thus

$$|A \cup B| = |A| + |B| - |A \cap B|. \tag{5.1}$$

Here the *inclusion-exclusion principle* is at work. First too many are *included*, but thereafter *excluded*.

Example 5.1. $A = \{1, 2, 3\}$ and $B = \{2, 3, 4\}$. $|A| = |B| = 3$, $|A \cap B| = 2$, so that $|A \cup B| = 3 + 3 - 2 = 4$. This is correct since $A \cup B = \{1, 2, 3, 4\}$.

What happens with 3 sets A, B, C (see Fig. 5.2)? In evaluating $|A \cup B \cup C|$, start off with $|A| + |B| + |C|$. Any element in both A and

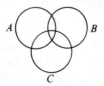

FIG. 5.2

B, or in both B and C, or in both C and A is included more than once. So the next attempt at a solution to consider is

$$|A| + |B| + |C| - |A \cap B| - |B \cap C| - |C \cap A|.$$

Not even this is correct, for if there are any elements in all three sets A, B, C, then they will have been included thrice and excluded thrice, and so must be added in once again. Thus

$$|A \cup B \cup C| = |A| + |B| + |C| - |A \cap B| - |B \cap C| - |C \cap A|$$

$$+ |A \cap B \cap C|. \tag{5.2}$$

The reader will now be able to deduce from the patterns in (5.1) and (5.2) a similar expression for $|A \cup B \cup C \cup D|$. Indeed,

$$|A \cup B \cup C \cup D| = |A| + |B| + |C| + |D| - |A \cap B| - |B \cap C|$$

$$- |C \cap D| - |D \cap A| - |A \cap C| - |B \cap D|$$

$$+ |A \cap B \cap C| + |A \cap B \cap D| + |A \cap C \cap D|$$

$$+ |B \cap C \cap D| - |A \cap B \cap C \cap D|.$$

All these expressions illustrate the basic inclusion-exclusion principle, which is now presented in a slightly different way.

Suppose that a collection of objects is given, along with a list of r properties which the objects may or may not possess, and suppose that it is required to find the number of objects which possess at least one of the properties. In the examples above, the first property was that of belonging to the set A, the second was that of belonging to B, and so on. Denote by $N(i, j, \ldots, k)$ the number of objects which possess each of the ith, jth, $\ldots$, kth properties (and possibly some others as well). Then the number of objects possessing at least one of the properties is

$$N(1) + N(2) + N(3) + \cdots + N(r) -$$

$$- \{N(1, 2) + N(1, 3) + N(2, 3) + \cdots + N(r - 1, r)\} +$$

$$+ \{N(1, 2, 3) + N(1, 2, 4) + \cdots + N(r - 2, r - 1, r)\} -$$

$$- \cdots$$

$$\vdots$$

$$+ (-1)^{r-1} N(1, 2, \ldots, r). \tag{5.3}$$

This result is perhaps more useful in its complementary form. Instead of asking how many objects possess at least one of the properties, it is asked how many possess *none* of the properties. Clearly this is obtained by subtracting the expression (5.3) from the total number of objects.

Proof of the principle (5.3). If an object possesses none of the r properties, then it clearly contributes nothing to (5.3). If an object possesses $t \geqslant 1$ properties, it must be shown that it contributes 1 to (5.3). But its contribution is

$$t - \binom{t}{2} + \binom{t}{3} - \binom{t}{4} + \cdots$$

$$= 1 - \left\{1 - t + \binom{t}{2} - \binom{t}{3} + \cdots\right\}$$

$$= 1 - (1 - 1)^t = 1.$$

Example 5.2. *Derangements.* The formula (4.22) for a_n, the number of derangements of n symbols, has the inclusion–exclusion look about it. Its appearance suggests that an alternative derivation is possible, and this is now confirmed. As objects, take the $n!$ possible permutations of the n symbols. An object possesses the ith property if the ith symbol appears in it in the ith place. Then the number of derangements is just the number of objects possessing none of the properties.

Using the notation of (5.3),

$$N(i) = (n - 1)!,$$

since the ith symbol is fixed and the remaining $(n - 1)$ can undergo any permutation. Similarly,

$$N(i, j) = (n - 2)!,$$

since two symbols are fixed, leaving $n - 2$ to be permuted;

$$N(i, j, k) = (n - 3)!,$$

and so on. Further, the number of terms of type $N(i)$ is $\binom{n}{1}$, of type $N(i, j)$ is $\binom{n}{2}$, and so on. Thus the number of permutations which satisfy at least one of the properties, i.e. which are *not* derangements, is

$$(n - 1)!\binom{n}{1} - (n - 2)!\binom{n}{2} + (n - 3)!\binom{n}{3} - \cdots.$$

The number of derangements is $n!$ minus this number, i.e.

$$n! - \frac{(n-1)! \, n!}{(n-1)! \, 1!} + \frac{(n-2)! \, n!}{(n-2)! \, 2!} - \frac{(n-3)! \, n!}{(n-3)! \, 3!} + \cdots$$

$$= n! \left\{ 1 - \frac{1}{1!} + \frac{1}{2!} - \frac{1}{3!} + \cdots + (-1)^n \frac{1}{n!} \right\},$$

which agrees with (4.22).

It so happens that it is profitable to consider this problem geometrically. Take an $n \times n$ chessboard (see Fig. 5.3), and represent a permutation of the numbers $1, 2, \ldots, n$ by placing a chesspiece on the square of the ith row and the jth column if the number i is permuted to the jth position. For example, the permutation 2413 is represented by the accompanying diagram (where the top row is taken as the first row, and the left column as the first column).

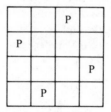

FIG. 5.3

Clearly a permutation corresponds to placing n pieces on an $n \times n$ board so that no two pieces lie in the same row or column. For a derangement, no piece must lie on the main diagonal (i.e. the diagonal from the top left to the bottom right). Thus formula (4.22) can be interpreted as giving the number of ways of placing n rooks on an $n \times n$ chessboard, with none on the main diagonal, so that no rook can taken any other rook. For, as is well known, a rook can only move along rows or columns.

This idea will be returned to later, but meanwhile another interpretation of (4.22) is given. Suppose that in constructing an $n \times n$ Latin square the numbers $1, 2, \ldots, n$ have been placed in some order in the first row. Then (4.22) gives the number of ways of choosing a second row for the square. This raises the question: assuming that the first $(r-1)$ rows have been chosen, can anything be said about the number of choices for the rth row? This is clearly closely related to the enumeration of permutations of $1, \ldots, n$ where there are $(r - 1)$ forbidden places for each number.

Exercises 5.1

1. Exam scripts of n students are returned to the class at random, one to each student. Show that the probability that no student receives his own script tends to $1/e$ as $n \to \infty$. (Probability = number of ways this can happen divided by the total number of ways in which the scripts can be returned.)

2. Each of a class of 50 students reads at least one of mathematics and physics. 30 read mathematics and 27 read both. How many read physics?

3. How many integers from 1 to 1000 are divisible by none of 3, 7, 11?

4. A survey carried out over a large number of citizens of a certain city revealed that 90 per cent of all people detest at least one of the pop stars Hairy, Dirty, and Screamer. 45 per cent detest Hairy, 28 per cent detest Dirty, and 46 per cent detest Screamer. If 27 per cent detest only Screamer, and 6 per cent detest all three, how many detest Hairy and Dirty but not Screamer?

5. Present the permutation 35142 by a chessboard diagram.

6. How many ways are there of placing 5 non-taking rooks on a 5 x 5 board? How many ways if none lie on the main diagonal? How many if exactly one lies on the main diagonal?

7. How many permutations are there of the digits $1, 2, \ldots, 8$ in which none of the patterns 12, 34, 56, 78 appears?

5.2. Rook polynomials

It has already been pointed out that the problem of derangements is equivalent to that of placing non-taking rooks on certain allowable squares of the chessboard. This suggests that some combinatorial problems may reduce to placing non-taking rooks on boards of various shapes and sizes.

Let C be an arbitrary board of any shape, with m squares. For each $k \leqslant m$, let $r_k(C)$ denote the number of ways of placing k non-taking rooks on C. Then the generating function for the numbers $r_k(C)$,

$$R(x, C) = r_0(C) + r_1(C)x + r_2(C)x^2 + \cdots + r_m(C)x^m,$$

is called the *rook polynomial* of the board C.

Example 5.3. Find the rook polynomial for an ordinary 4 x 4 board.

Solution. The numbers $r_i(C)$, $i = 0, \ldots, 16$ must be evaluated. Clearly $r_i(C) = 0$ for all $i > 4$.

$r_0(C)$ = number of ways of placing no non-taking rooks on C, = 1
(the only way being to leave the board empty).

$r_1(C) = 16$, since there are 16 squares to choose from. Next, $r_2(C)$ is the number of ways of placing two non-taking rooks on C. These rooks must lie in different rows and columns. The number of ways of choosing two rows in $\binom{4}{2}$. Once the rows are chosen, a rook can be placed in the first one in any of 4 ways, and another rook in the second row in any of 3 places.

Thus $\qquad\qquad r_2(C) = \binom{4}{2} \cdot 4 \cdot 3 = 72$. Similarly

$$r_3(C) = \binom{4}{3} \cdot 4 \cdot 3 \cdot 2 = 96, r_4(C) = \binom{4}{4} \cdot 4 \cdot 3 \cdot 2 \cdot 1 = 24.$$

Thus

$$R(x, C) = 1 + 16x + 72x^2 + 96x^3 + 24x^4.$$

The reader should now try Exercises 5.2, question 1.

Faced with a more awkwardly shaped board, the problem of finding the rook polynomial would prove to be near impossible if it were not that some tricks exist whereby a board can be reduced to a simpler one. One such trick is concerned with boards which fall into two or more non-interfering parts. Two parts A, B of a chessboard C are *non-interfering* if no square in A is in the same row or column of C as any square of B. The board in Fig. 5.4 falls into 3 non-interfering parts.

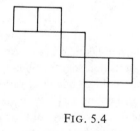

FIG. 5.4

PROPERTY 1. *If a chessboard C consists of two non-interfering parts, then the rook polynomial for C is just the product of the rook polynomials for the parts A and B.*

Proof. When k non-taking rooks are placed on C, t will be placed on A and $(k - t)$ on B, for some t, $0 \leqslant t \leqslant k$. Since the $r_t(A)$ possible placings of t rooks on A can each occur along with any of the $r_{k-t}(B)$ placings on B (for A and B do not interfere with one another), it follows that

$$r_k(C) = r_0(A)r_k(B) + r_1(A)r_{k-1}(B) + \cdots + r_k(A)r_0(B).$$

But the expression on the right is simply the coefficient of x^k in

$$\{r_0(A) + r_1(A)x + r_2(A)x^2 + \cdots\}\{r_0(B) + r_1(B)x + r_2(B)x^2 + \cdots\},$$

i.e. in the product of the rook polynomials for A and B.

Example 5.4. Suppose that C consists of n non-interfering 2 x 2 blocks (Fig. 5.5).

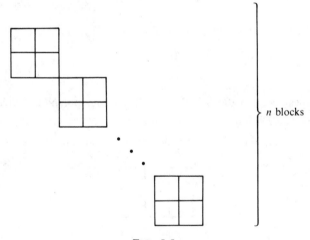

n blocks

FIG. 5.5

Solution. The rook polynomial for one block is $1 + 4x + 2x^2$. The rook polynomial for C is therefore $(1 + 4x + 2x^2)^n$.

Property 1, although useful, is not widely applicable. The problem still remains of how to deal with a board which does not fall into non-interfering parts. The next property is of use here.

PROPERTY 2. *Given a chessboard C, choose any square of C and let D denote the board obtained by deleting from C every square in the same row or column as the chosen square (including the chosen square itself).*

Let E denote the board obtained from C by deleting only the chosen square. Then

$$R(x, C) = xR(x, D) + R(x, E).$$

Proof. If k non-taking rooks are placed on C, then the chosen square either is or is not used. If it is used, then $(k - 1)$ rooks are left to be placed on D, and this can be done in $r_{k-1}(D)$ ways. If it is not used, then k rooks have to be placed on E, and this can be done in $r_k(E)$ ways. Thus

$$r_k(C) = r_{k-1}(D) + r_k(E),$$

so that

$$R(x, C) = \sum_{k=0}^{\infty} r_k(C)x^k$$

$$= \sum_{k=0}^{\infty} r_{k-1}(D)x^k + \sum_{k=0}^{\infty} r_k(E)x^k$$

$$= xR(x, D) + R(x, E).$$

By repeated applications of Property 2, the rook polynomial of any board can be found.

Example 5.5. Find the rook polynomial of the board of Fig. 5.6.

FIG. 5.6

Solution. Choosing the centre square,

$$R(x, C) = xR(x, D) + R(x, E), \tag{5.4}$$

where

$$D = \square \ \square \quad \text{and} \quad E = \boxed{}.$$

Now,

$$R(x, D) = 1 + 2x \tag{5.5}$$

and, by Property 1,

$$R(x, E) = R(x, H)R(x, K),$$

where

$$H = \boxminus\boxminus \quad \text{and} \quad K = \square .$$

Since $R(x, H) = 1 + 4x + 2x^2$ and $R(x, K) = 1 + x$, it follows that

$$R(x, E) = (1 + x)(1 + 4x + 2x^2). \tag{5.6}$$

From (5.4)–(5.6) it now follows that

$$R(x, C) = x(1 + 2x) + (1 + x)(1 + 4x + 2x^2)$$
$$= 1 + 6x + 8x^2 + 2x^3.$$

This whole argument can be written more clearly in the following symbolic way:

$$R(x, C) = xR(\square\ \square) + R\left(\boxplus\boxplus\right)$$
$$= x(1 + 2x) + R\left(\boxminus\boxminus\right)R\left(\square\right)$$
$$= x(1 + 2x) + (1 + 4x + 2x^2)(1 + x).$$

This abbreviated notation is used in the next example.

Example 5.6. $R\left(\boxplus\right) = xR(\square) + R\left(\blacksquare\square\right)$

$$= x(1 + x) + xR\left(\begin{smallmatrix}\square\\\square\end{smallmatrix}\right) + R\left(\text{board}\right)$$
$$= x(1 + x) + x(1 + 2x) + R\left(\boxplus\right)R(\square)$$
$$= x(2 + 3x) + (1 + 3x + x^2)(1 + x)$$
$$= 1 + 6x + 7x^2 + x^3.$$

Applications of rook polynomials

Example 5.7. The manager of a firm has 5 employees to be assigned to 5 different jobs. The men are A, B, C, D, E and the jobs are a, b, c, d, e. He considers that A is unsuited for jobs b and c, B unsuited for a and c, C unsuited for b, d and e, D suited for all and E unsuited for d. In how many ways can he assign the jobs to men suited to them?

Solution. The board shown in Fig. 5.7 represents the situation. The problem is to find the coefficient of x^5 in the rook polynomial for this board. At this point the reader will probably hold back at the mere thought of finding the rook polynomial, due to the amount of work

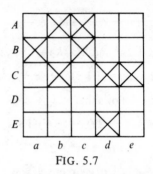

FIG. 5.7

involved. In fact, it would be much easier to find the rook polynomial for the board consisting of the forbidden positions. This polynomial will now be found before its usefulness to the original problem is explained.

$$R\left(\begin{array}{c}\blacksquare\end{array}\right) = xR\left(\begin{array}{c}\end{array}\right) + R\left(\begin{array}{c}\blacksquare\end{array}\right)$$

$$= x\left\{xR(\square) + R\left(\begin{array}{c}\end{array}\right)\right\} + R(\square)R\left(\begin{array}{c}\blacksquare\end{array}\right)$$

$$= x\{x(1+x) + 1 + 3x + 2x^2\} + (1+x)\left\{xR\left(\begin{array}{c}\end{array}\right)\right.$$

$$\left. + R\left(\begin{array}{c}\end{array}\right)\right\}$$

$$= x(1 + 4x + 3x^2) + (1+x)\left\{x(1 + 3x + x^2) + \right.$$

$$\left. + (1+x)R\left(\begin{array}{c}\end{array}\right)\right\}$$

$$= (x + 4x^2 + 3x^3) + (1+x)\{x + 3x^2 + x^3 +$$

$$+ (1+x)(1 + 4x + 3x^2)\}$$

$$= 1 + 8x + 20x^2 + 17x^3 + 4x^4. \tag{5.7}$$

This is the rook polynomial for the board consisting of the forbidden squares. Now the assignment of jobs to men can be considered as permutations of the numbers $1, \ldots, 5$. For example, if A gets job c, B gets

d, C gets b, D gets a, and E gets e, the assignment corresponds to the permutation 34215, since, for example, the first man gets the third job and the fourth man gets the first job. The key to the problem now lies in the following theorem.

THEOREM 5.1. *The number of permutations of n symbols in which no symbol is in a forbidden position is*

$$\sum_{k=0}^{n} (-1)^{k}(n - k)! \, r_k$$

where r_k is the number of ways of placing k non-taking rooks on the board of forbidden positions.

Solution to Example 5.7 (continued). Assuming for the moment that the theorem has been proved, and noting that, from (5.7),

$$r_0 = 1, r_1 = 8, r_2 = 20, r_3 = 17, r_4 = 4,$$

and $n = 5$, the number of ways of assigning the jobs to the men is

$$5! - 4! \, 8 + 3! \, 20 - 2! \, 17 + 4 = 18.$$

Thus a knowledge of the rook polynomial for the board of forbidden squares leads very quickly to information about the permitted squares.

Proof of Theorem 5.1. In the notation of the inclusion-exclusion principle, suppose that a permutation possesses the ith property if the ith symbol is in a forbidden position. Then the number of permutations with no symbol in a forbidden position is

$$n! - \{N(1) + \cdots + N(n)\} + \{N(1, 2) + \cdots\} - \cdots.$$

Now each $N(i)$ is equal to $s_i(n - 1)!$ where s_i is the number of forbidden squares in the ith row, since the ith symbol can be placed on any of these s_i squares and the remaining symbols can be placed in $(n - 1)!$ ways. Since $s_1 + \cdots + s_n = r_1$, it follows that

$$N(1) + \cdots + N(n) = (n - 1)! \, (s_1 + \cdots + s_n) = (n - 1)! \, r_1.$$

Similarly,

$$N(1, 2) + N(1, 3) + \cdots + N(n - 1, n) = (n - 2)! \, r_2$$

and so on.

Example 5.8. In constructing a 6 × 6 Latin square, the first two rows have been chosen as follows:

$$1 \quad 2 \quad 3 \quad 4 \quad 5 \quad 6$$
$$2 \quad 4 \quad 1 \quad 3 \quad 6 \quad 5.$$

By Hall's theorem (3.3) it is definitely possible to find a suitable third row. But how many possibilities are there?

Solution. The problem is: how many permutations of $1, \ldots, 6$ are there with no symbol in a forbidden position, the forbidden positions being represented by crosses in the diagram (Fig. 5.8)?

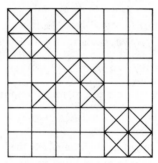

FIG. 5.8

Following the method of the previous example, the first thing to do is to obtain the rook polynomial for the board of forbidden positions. This is

$$R\left(\text{▨}\right)R\left(\text{▨}\right) = (2x^4 + 16x^3 + 20x^2 + 8x + 1)(1 + 4x + 2x^2)$$

$$= (4x^6 + 40x^5 + 106x^4 + 112x^3 + 54x^2 + 12x + 1).$$

In the notation of Theorem 5.1, $r_6 = 4, r_5 = 40, r_4 = 106, r_3 = 112,$ $r_2 = 54, r_1 = 12,$ and $r_0 = 1,$ so that the number of possibilities for the third row of the square is

$$6! - 12.5! + 54.4! - 112.3! + 106.2! - 40 + 4 = 70.$$

Exercises 5.2

1. Find the rook polynomial for an ordinary 8 × 8 chessboard.

2. Find the rook polynomials for the following boards:

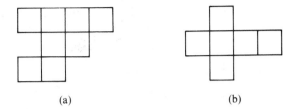

(a)   (b)

3. A six-a-side football team is to consist of the players $A, B, \ldots, F$. A refuses to play in positions 1 or 2, B in position 4, C in positions 1 or 5, D in 2, E in 4, and F in 4 or 6. How many ways are there of assigning agreeable positions to the six players?

4. The first two rows of a 5 x 5 Latin square are 1, 2, 3, 4, 5 and 2, 3, 4, 5, 1. In how many ways can a third row be chosen?

5. Deduce (4.22) from Theorem 5.1 by choosing a suitable board.

6. Show that the rook polynomial $R_{n,m}(x)$ for a rectangular $n \times m$ board satisfies the recurrence relation

$$R_{n,m}(x) = R_{n-1,m}(x) + mx R_{n-1,m-1}(x).$$

7. A computer matching service has five male subscribers A, B, C, D, E and four female subscribers a, b, c, d. After analysing their interests and personalities, the computer decides that a is unsuitable for B and C, b unsuitable for C, c for A and E, d for B. In how many ways can the female subscribers be matched?

8. By trial and error, verify that there are four possible third rows for a Latin square whose first two rows are 1, 2, 3, 4 and 2, 1, 4, 3, whereas there are only two possibilities if the first two rows are 1, 2, 3, 4 and 2, 4, 1, 3. Thus, although the number of choices for the next row is always $\geqslant 1$, the actual number of choices depends on the choice of the previous rows.

9. In the light of question 8, repeat question 4 if the second row is 2, 3, 1, 5, 4.

10. Let $L_k(x)$ denote the rook polynomial of the board consisting of the first k crossed squares (reading down and to the right) of the following $n \times n$ board.

Prove that $L_0(x) = 1$, $L_1(x) = 1 + x$, and that, for $k \geqslant 2$,

$$L_k(x) = L_{k-1}(x) + x L_{k-2}(x).$$

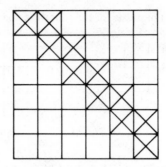

Using Exercises 4.4, question 5, deduce that the rook polynomial of the board consisting of all the crossed squares is

$$L_{2n-1}(x) = \sum_r \binom{2n-r}{r} x^r.$$

11. Let $R_n(x)$ denote the rook polynomial of the board of crosses obtained by adding a cross to the bottom left-hand square of the above board. Show that

$$R_n(x) = L_{2n-1}(x) + xL_{2n-3}(x)$$

$$= \sum_r \frac{2n}{2n-r} \binom{2n-r}{r} x^r.$$

12. The '*problème des ménages*'. Let a_n denote the number of arrangements of n pairs of husbands and wives round a circular table with men and women in alternate places and no man sitting beside his wife. Show that $a_n = (n-1)! \, b_n$ where b_n is the number of ways of placing n non-taking rooks on the board of non-crossed squares of the preceding question. Using Theorem 5.1, deduce that

$$b_n = \sum_k (-1)^k \frac{2n}{2n-k} \binom{2n-k}{k} (n-k)!.$$

Verify this in the case $n = 5$ (see question 4).

6 Block designs and error-correcting codes

6.1. Block designs

THE origins of the theory of block designs can be traced back to the problem of designing certain types of statistical experiment. It is therefore not insignificant that the name of the distinguished statistician Fisher is attached to one of the first results in the subject (Theorem 6.2).

The idea behind a block design can be seen in the following type of problem. Suppose that a number of brands of instant coffee are to be tested among a number of housewives, the object of the experiment being to let the ladies compare the different brands and decide on their relative merits. To make the tests as fair as possible, it is decided that the following conditions should be satisfied:

(1) each housewife should taste the same number of brands;

(2) each pair of brands should be compared by the same number of housewives.

Clearly, one way of achieving this would be to give every housewife every brand of coffee, but this is wasteful and time consuming. The problem is to achieve the aim more economically. Mathematically, all that is involved is a set S of *varieties* (the brands of coffee), and a collection of subsets of S (each subset consisting of those varieties which a particular housewife tastes) called *blocks*, with the properties:

(a) each block has the same number of elements;

(b) every pair of varieties is contained in the same number of blocks.

DEFINITION 6.1. A *block design* is a family of b subsets of a set S of v elements such that, for some fixed k and λ,

(1) each subset has k elements,

(2) each pair of elements of S occur together in exactly λ subsets.

The elements of S are called the *varieties*, and the subsets of S are called the *blocks*.

Example 6.1. Take $S = \{1, 2, \ldots, 7\}$, and consider the following seven subsets of S:

$$\{1, 2, 4\}, \{2, 3, 5\}, \{3, 4, 6\}, \{4, 5, 7\}, \{5, 6, 1\}, \{6, 7, 2\}, \{7, 1, 3\}.$$

Here $b = 7, v = 7, k = 3, \lambda = 1$. To see that $\lambda = 1$, consider any pair of elements, say 4 and 6, and verify that exactly one of the seven subsets contains both 4 and 6. Do this for each pair.

This design could be used to compare 7 brands of coffee, using seven housewives. Each housewife is given 3 brands, and any particular pair of brands will be compared by exactly one housewife.

There is a simple geometrical representation of the above design. The elements 1, . . ., 7 are represented by points, and the blocks are represented by lines (all but one being a straight line). This is the simplest example of a *finite projective plane*, where the elements are usually called points and the blocks are called lines. This one is known as the *seven-point plane* (Fig. 6.1).

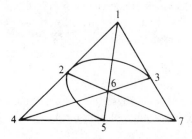

FIG. 6.1. The seven-point plane.

But there is another useful way of representing the design of Example 6.1. The first set {1, 2, 4} can be represented by the following string of 0s and 1s:

$$1\ 1\ 0\ 1\ 0\ 0\ 0.$$

There is a 1 in the first, second, and fourth places because the set consists of the first, second, and fourth elements. Similarly, {2, 3, 5} can be represented by

$$0\ 1\ 1\ 0\ 1\ 0\ 0.$$

Representing each set in this way, and listing the strings one under the other, the following matrix, called the *incidence matrix* of the design, is formed

$$\begin{bmatrix} 1 & 1 & 0 & 1 & 0 & 0 & 0 \\ 0 & 1 & 1 & 0 & 1 & 0 & 0 \\ 0 & 0 & 1 & 1 & 0 & 1 & 0 \\ 0 & 0 & 0 & 1 & 1 & 0 & 1 \\ 1 & 0 & 0 & 0 & 1 & 1 & 0 \\ 0 & 1 & 0 & 0 & 0 & 1 & 1 \\ 1 & 0 & 1 & 0 & 0 & 0 & 1 \end{bmatrix}.$$

Each row represents a subset or block, and each column gives information about a particular element or variety. For example, by looking at the third column it is deduced that the element 3 occurs in the second, third, and seventh sets. The condition that any pair of elements occur together in exactly one block is represented by the property that any two columns both have 1s in the same row exactly once. For example, the first and seventh columns both have a 1 in the seventh row; this means that the elements 1 and 7 occur together only in the seventh set.

The advantage of using incidence matrices to describe a block design instead of listing the sets element by element is that the structure of the design is seen more clearly without any irrelevant information such as the names of the elements confusing the issue. It is also easier to scan the columns to find how many sets contain a given element than to look through a list of sets. Note also that the number of rows is b, and the number of columns is v.

The conditions for a block design imply a further condition, namely that each variety must occur in the same number of blocks. The proof of this acts as an introduction to simple but important counting ideas which will be much used in the next chapter.

THEOREM 6.1. *In a block design each element lies in exactly r blocks, where*

$$r(k - 1) = \lambda(v - 1) \quad and \quad bk = vr. \tag{6.1}$$

Proof. Concentrate on any one of the elements, and suppose that it occurs in r blocks, for some r. Each of these r blocks contains $(k - 1)$ other elements, so that the number of pairs including this chosen element is $r(k - 1)$. But there are $(v - 1)$ elements with which it can be paired, and each pair occurs λ times. Hence $r(k - 1) = \lambda(v - 1)$. Since k, v, and λ are fixed, it follows that r must be the same for each element. For this fixed value of r, each element therefore has r appearances in the blocks, so that there are vr appearances of elements altogether. But there are b blocks each with k elements, so the number of appearances must also be bk. Thus $bk = rv$.

The five parameters b, v, r, k, λ of a block design are therefore not independent, but have two restrictions as stated in the theorem. Often a block design is referred to as a (b, v, r, k, λ)-*configuration*; for example, the seven-point plane is a $(7, 7, 3, 3, 1)$-configuration. Whatever b, v, r, k, λ are, they must satisfy (6.1), but conversely, if five numbers b, v, r, k, λ satisfy (6.1), there is no guarantee that a (b, v, r, k, λ)-configuration exists. For example, it is known that a finite projective plane with $b = v = 43$, $r = k = 7$, and $\lambda = 1$ does not exist (see Theorem 6.4).

The seven-point plane has a further property which is not possessed by all block designs, namely that $b = v$. This means that the number of blocks is the same as the number of elements, so that the incidence matrix is a square matrix. Such designs are called *square* or *symmetric designs* although the second description is misleading since the incidence matrix need not be symmetric about the main diagonal. The reason for the name will appear later. Since $b = v$ implies $r = k$, square designs are completely determined by the three parameters v, k, λ and hence are often called (v, k, λ)-configurations. The seven-point plane is a $(7, 3, 1)$-configuration. Condition (6.1) becomes

$$k(k - 1) = \lambda(v - 1). \qquad (6.2)$$

Equality of b and v is in a sense the extreme case since b can never be smaller than v in a block design. This is Fisher's result, proved in 1940.

THEOREM 6.2 (Fisher). *For a (b, v, r, k, λ)-configuration,*

$$b \geqslant v.$$

Proof. Let A be the incidence matrix, so that A has b rows and v columns. The key idea in the proof is to determine the matrix $C = A'A$. Here A' is the transposed matrix of A, obtained by writing the rows of A as columns, and the columns as rows. The element a'_{ij} in the ith row and jth column of A' is equal to a_{ji}, the element in the jth row and the ith column of A. Then $C = (c_{ij})_{v \times v}$, where

$$c_{ij} = \sum_h a'_{ih} a_{hj}$$

$$= \sum_h a_{hi} a_{hj}.$$

In particular,

$$c_{ii} = \sum_h a_{hi}^2 = \sum_h a_{hi}$$

since each a_{ij} is 0 or 1 and $0^2 = 0$ and $1^2 = 1$. But $a_{hi} = 1$ if and only if the ith element is in the hth set, and is 0 otherwise. Thus

$$c_{ii} = \sum_h a_{hi} = \text{number of sets containing the } i\text{th element}$$

$$= r.$$

Also, if $i \neq j$,

$$c_{ij} = \sum_h a_{hi} a_{hj}.$$

Now $a_{hi} a_{hj}$ is equal to 1 if and only if $a_{hi} = a_{hj} = 1$, i.e. only if the hth set contains both the ith and the jth elements. There are λ such hs. Thus $c_{ij} = \lambda$, and

$$C = A'A = \begin{bmatrix} r & \lambda & \lambda & \dots & \lambda \\ \lambda & r & \lambda & \dots & \lambda \\ \vdots & & & & \vdots \\ \lambda & \lambda & \lambda & \dots & r \end{bmatrix}.$$

If I is used to denote the unit matrix with 1s down the main diagonal and 0s elsewhere, and J denotes the matrix with every entry equal to 1, this result can be written as

$$A'A = (r - \lambda)I + \lambda J. \tag{6.3}$$

Exercise for the reader. If J is $v \times v$, then $J^2 = vJ$.

To prove that $b \geqslant v$, it will be assumed that $b < v$ and a contradiction will be obtained. So suppose that $b < v$. A will then have fewer rows than columns, so add a further $(v - b)$ rows to A, all of whose elements are 0s. The matrix A will then be transformed into a new matrix A_1, but A_1 will also have the property that

$$A_1'A_1 = (r - \lambda)I + \lambda J.$$

Since A_1 has at least one row of 0s, its determinant must be zero. Thus

$$\det(A_1'A_1) = \det(A_1') \cdot \det(A_1) = 0. \tag{6.4}$$

However,

$$
\det
\begin{bmatrix}
r & \lambda & \lambda & \ldots & \lambda \\
\lambda & r & \lambda & \ldots & \lambda \\
\vdots & & & & \\
\lambda & \lambda & \lambda & \ldots & r
\end{bmatrix}
= \det
\begin{bmatrix}
r & \lambda & \lambda & \ldots & \lambda \\
\lambda - r & r - \lambda & 0 & \ldots & 0 \\
\lambda - r & 0 & r - \lambda & \ldots & 0 \\
\vdots & & & & \\
\lambda - r & 0 & 0 & \ldots & r - \lambda
\end{bmatrix}
$$

on substracting the first row from each of the others, and this, in turn, on adding to the first column all the other columns, is equal to

$$
\det
\begin{bmatrix}
r + \lambda(v - 1) & \lambda & \lambda & \ldots & \lambda \\
0 & r - \lambda & 0 & \ldots & 0 \\
0 & 0 & r - \lambda & \ldots & 0 \\
\vdots & & & & \\
0 & 0 & 0 & \ldots & r - \lambda
\end{bmatrix}
$$

$$= \{r + (v - 1)\lambda\}.(r - \lambda)^{v-1}$$

$$= rk(r - \lambda)^{v-1}$$

$$\neq 0, \quad \text{since (6.1) implies that } r > \lambda.$$

This contradicts (6.4) and so completes the proof of the theorem.

Exercises 6.1

1. The following 12 sets form a (b, v, r, k, λ)-configuration.

$$
\begin{array}{cccc}
\{1, 2, 3\} & \{4, 5, 6\} & \{7, 8, 9\} & \{1, 4, 7\} \\
\{2, 5, 8\} & \{3, 6, 9\} & \{1, 5, 9\} & \{2, 6, 7\} \\
\{3, 4, 8\} & \{1, 6, 8\} & \{2, 4, 9\} & \{3, 5, 7\}
\end{array}
$$

Write down the incidence matrix and check that $b = 12, v = 9, r = 4$, $k = 3, \lambda = 1$. Verify that $bk = rv$ and $r(k - 1) = \lambda(v - 1)$. Explain how this design could be used to test 9 detergents with the help of 12 housewives, or with the help of 3 housewives on 4 consecutive days.

2. Find $A'A$ and AA' for the seven-point plane.

3. Show that there exists no $(12, 8, 3, 2, 1)$-configuration.

4. Define the *complement* of a design D to be the design obtained by changing 0 to 1 and 1 to 0 throughout the incidence matrix of D. If D is a (b, v, r, k, λ)-configuration, show that its complement is a $(b, v, b - r, v - k, b - 2r + \lambda)$-configuration.

5. Derive from question 1 above a $(12, 9, 8, 6, 5)$-configuration.

6.2. Square block designs

In the special case of a square design, (6.3) becomes

$$A'A = (k - \lambda)I + \lambda J = \begin{bmatrix} k & \lambda & \lambda & \dots & \lambda \\ \lambda & k & \lambda & \dots & \lambda \\ \vdots & & & & \vdots \\ \lambda & \lambda & \lambda & \dots & k \end{bmatrix}. \quad (6.5)$$

As has been pointed out already, the incidence matrix A of a (v, k, λ)-configuration need not itself be symmetric. The reason for calling a square block design symmetric is that there is the following symmetry in the properties of the rows and columns of the incidence matrix:

(1) Any row contains k 1s.

(2) Any column contains k 1s.

(3) Any pair of columns both have 1s in exactly λ rows.

(4) Any pair of rows both have 1s in exactly λ columns.

Property (4) has not yet been proved, but it will be shown to follow from (1), (2), and (3).

Properties (2) and (3) are contained in (in fact are equivalent to) the statement (6.5), whereas (1) and (4) can together be expressed as

$$AA' = (k - \lambda)I + \lambda J. \quad (6.6)$$

It will be shown that (2) and (3) together imply (1) and (4), and, conversely, (1) and (4) together will imply (2) and (3). Once this has been established, it will follow that A is the incidence matrix of a (v, k, λ)-configuration if A satisfies either (6.5) or (6.6).

THEOREM 6.3. *If A is a square $(0, 1)$-matrix (i.e. a matrix all of whose entries are 0 or 1) and if A satisfies (6.5) with $k > \lambda$, then (6.6) also holds.*

Proof. Since the diagonal elements of $A'A$ are all k, each column of A contains exactly k 1s. Thus

$$JA = kJ.$$

It follows that $JAA^{-1} = kJA^{-1}$, whence $JA^{-1} = \frac{1}{k}J$. Thus

$$AA' = AA'AA^{-1} = A\{(k - \lambda)I + \lambda J\}A^{-1}$$

$$= (k - \lambda)I + \lambda AJA^{-1}$$

$$= (k - \lambda)I + \lambda k^{-1}AJ.$$

The result will therefore follow on showing that $AJ = kJ$. This will certainly be true if property (1) above is assumed, so it is now confirmed that (1), (2), and (3) imply (4).

If $AJ = kJ$ is not assumed, it must be proved. From (6.5),

$$A' = (k - \lambda)A^{-1} + \lambda JA^{-1}$$
$$= (k - \lambda)A^{-1} + \lambda k^{-1}J.$$

Thus

$$JA' = (k - \lambda)JA^{-1} + \lambda k^{-1}vJ = \{(k - \lambda)k^{-1} + \lambda k^{-1}v\}J, \qquad (6.7)$$

whence, on transposing both sides,

$$AJ = k^{-1}(k - \lambda + \lambda v)J.$$

Once again, the proof would be complete if it were known that $k - \lambda + \lambda v = k^2$. This is known if A is the incidence matrix of a (v, k, λ)-configuration. To prove the theorem in complete generality, it therefore remains to *prove* that $k - \lambda + \lambda v = k^2$. The trick is to consider $JA'J$.

First, $JA = kJ$, so $A'J = kJ$, from which $JA'J = kJ^2 = kvJ$. Secondly, by (6.7),

$$JA'J = k^{-1}(k - \lambda + \lambda v)J^2 = k^{-1}v(k - \lambda + \lambda v)J.$$

Comparing these two expressions it follows that $k^2 = k - \lambda + \lambda v$.

Example 6.2. A *finite projective plane of order n* is defined to be a (v, k, λ)-configuration for which $v = n^2 + n + 1$, $k = n + 1$, and $\lambda = 1$, for some positive integer $n \geqslant 2$. The seven-point plane corresponds to $n = 2$. In a plane of order n there are therefore $(n^2 + n + 1)$ points and $(n^2 + n + 1)$ lines, and the four properties listed on p. 83 become as follows.

(1) Any line contains $(n + 1)$ points.
(2) Any point lies on $(n + 1)$ lines.
(3) Any pair of points are joined by exactly one line.
(4) Any pair of lines intersect in exactly one point.

These four properties can be checked for $n = 2$ by studying Fig. 6.1. The next plane, corresponding to $n = 3$, is a thirteen-point plane with 4 points on each line and 4 lines through each point. See the Exercises 6.4 for its construction.

The major unsolved problem for finite projective planes is to find all those values of n for which a plane of order n exists. The following statements sum up the state of present knowledge.

(a) A plane of order n definitely exists if $n \geqslant 2$ is a prime or a power of a prime.

(b) No plane of any other order is known to exist.

(c) There is definitely no plane of order 6, or in general of any order n, where n is of the form $(4k + 1)$ or $(4k + 2)$, and is divisible an odd number of times by a prime of the form $(4h + 3)$.

The smallest values of n which are excluded by (c) are $n = 6, 14, 22$. The smallest number not covered by (a) and (c) is 10, and it is still not known whether or not a plane of order 10 exists:

Unsolved problem: Is it possible to construct a square $(0, 1)$-matrix A with 111 rows and 111 columns, each row and column containing exactly eleven 1s, such that

$$AA' = 10I + J.$$

The statement (c) above is due to two North American mathematicians Bruck and Ryser. Their proof is a delightful example of the ingenuity and cunning which abound in this branch of mathematics. The proof is accessible to anyone who has done a little matrix algebra, and is now presented in the simplest case of $n = 6$.

THEOREM 6.4. *There is no finite projective plane of order 6.* It will be convenient to note a few preliminary results before embarking on the proof of the theorem. In what follows, I_n will denote the $n \times n$ unit matrix.

LEMMA 1. *If H is the 4 × 4 matrix defined by*

$$H = \begin{bmatrix} 2 & 1 & 1 & 0 \\ 1 & -2 & 0 & -1 \\ 1 & 0 & -2 & 1 \\ 0 & 1 & -1 & -2 \end{bmatrix}$$

then $HH' = 6I_4$. (I_4 denotes the 4 × 4 unit matrix.)

LEMMA 2. *There are no integers a, b, c such that $a^2 + b^2 = 6c^2$, apart from $a = b = c = 0$.*

Proof of Lemma 2. Suppose such integers do exist. If a, b, c have a common factor it can be divided out, so it can be assumed that no positive integer > 1 divides each of a, b, c. Now $6c^2$ is divisible by 3, so $a^2 + b^2$ must also be divisible by 3. The reader should be able to check that a sum

$a^2 + b^2$ can only be divisible by 3 if both a and b are divisible by 3. But then a^2 and b^2, and hence $a^2 + b^2$, are divisible by 9. This implies that $6c^2$ must be divisible by 9, i.e. $2c^2$ is divisible by 3. Thus 3 divides c^2 and hence also divides c. But to have a, b, c all divisible by 3 is a contradiction.

Proof of Theorem 6.4. Suppose there is a plane of order 6. Its incidence matrix A will have 43 rows and columns, and will satisfy

$$AA' = \begin{bmatrix} 7 & 1 & 1 & \ldots & 1 \\ 1 & 7 & 1 & \ldots & 1 \\ \vdots & & & & \\ 1 & 1 & 1 & \ldots & 7 \end{bmatrix}.$$

Thus, if

$$B = \left[\begin{array}{ccc:c} & & & 0 \\ & A & & \vdots \\ & & & 0 \\ \hdashline 0 & 0 \ldots 0 & & 1 \end{array} \right]_{44 \times 44}$$

then

$$BI_{44}B' = \begin{bmatrix} 7 & 1 & \ldots & 1 & 0 \\ 1 & 7 & & 1 & 0 \\ \vdots & & \ddots & \vdots & \vdots \\ 1 & 1 & \ldots & 7 & 0 \\ 0 & 0 & \ldots & 0 & 1 \end{bmatrix}. \tag{6.8}$$

Also, if H is as in Lemma 1, and if

$$K = \begin{bmatrix} H & & & & \\ & H & & \text{\Large 0} & \\ & & H & & \\ \text{\Large 0} & & & \ddots & \\ & & & & H \end{bmatrix}_{44 \times 44}$$

$$KI_{44}K' = 6I_{44}. \tag{6.9}$$

Now the quadratic form associated with the matrix on the right of (6.8) is

$$7(x_1^2 + \cdots + x_{43}^2) + x_{44}^2 + \sum_{i \neq j} x_i x_j$$

$$= (x_1 + \cdots + x_{43})^2 + x_{44}^2 + 6(x_1^2 + \cdots + x_{43}^2),$$

and the quadratic form associated with the matrix on the right of (6.9) is

$$6(x_1^2 + \cdots + x_{44}^2).$$

By (6.8) and (6.9), these forms are both transformed into the form associated with the matrix I_{44} by a non-singular linear change of variable. Combining these changes together, a non-singular change of variable

$$\begin{bmatrix} x_1 \\ \vdots \\ x_{44} \end{bmatrix} = P \begin{bmatrix} y_1 \\ \vdots \\ y_{44} \end{bmatrix}$$

must exist such that

$$6(y_1^2 + \cdots + y_{44}^2) = (x_1 + \cdots + x_{43})^2 + x_{44}^2 + 6(x_1^2 + \cdots + x_{43}^2),$$
(6.10)

the matrix P being non-singular and with rational numbers as its entries. In particular,

$$x_1 = p_{1,1}y_1 + \cdots + p_{1,44}y_{44}.$$
(6.11)

If $p_{1,1} = 1$, put $x_1 = y_1$. If $p_{1,1} = 1$, put $x_1 = -y_1$. In either case, $x_1^2 = y_1^2$ and, by (6.10), with x_1 replaced by y_1, y_1 now depends on $y_2, \ldots, y_{44}$. In the relation for x_2 corresponding to (6.11), y_1 can therefore be replaced to give

$$x_2 = q_2 y_2 + \cdots + q_{44} y_{44},$$

with each q_i rational. Now set $x_2 = \pm y_2$ as before. This induces a dependence relation expressing y_2 in terms of $y_3, \ldots, y_{44}$. Continue this process to reach eventually

$$x_{43} = r_{43} y_{43} + r_{44} y_{44}.$$

Now put $x_{43} = \pm y_{43}$ to get

$$y_{43} = gy_{44}$$

for some rational number g. So far the y_i have been unspecified. Choose y_{44} to be any non-zero rational. Then $y_{43}, \ldots, y_1$ are all uniquely specified, as are all the x_i. Moreover, $x_i^2 = y_i^2$ for each $i = 1, \ldots, 43$, so (6.10) becomes

$$6y_{44}^2 = x_{44}^2 + (x_1 + \cdots + x_{44})^2$$

Thus there is a rational solution of $6c^2 = a^2 + b^2$. On multiplying throughout by the square of the denominator, a contradiction to Lemma 2 is obtained.

Exercises 6.2

1. Why is it known in the proof of Theorem 6.3 that A^{-1} exists?

2. The blocks of a (v, k, λ)-configuration are the complements of the lines of the seven-point plane. Show that the design is a $(7, 4, 2)$-configuration and write down its incidence matrix.

3. Show that the proof of Theorem 6.4 cannot be adapted to deal with $n = 10$ by showing that the equation $a^2 + b^2 = 10c^2$ *does* have solutions in non-zero integers a, b, c.

4. Theorem 6.3 tells us that if A is the incidence matrix of a (v, k, λ)-configuration, then so is A'. The two configurations may not be the same, but are called the *duals* of one another. Write down the incidence matrix of the dual of the seven-point of Example 6.1 and, by reversing the orders of both the rows and the columns, show that the dual is the same as the original (apart from the labelling of the points and lines). The seven-point plane is therefore said to be *self-dual.*

6.3. Hadamard configurations

This section is intended as an introduction to a family of (v, k, λ)-configurations called *Hadamard configurations*, in which $v = 4m - 1$, $k = 2m - 1, \lambda = m - 1$ for some integer $m \geqslant 2$. The seven-point plane is such a design corresponding to $m = 2$.

One attractive but simple property of these designs is that once one such design is known, an infinite number of other Hadamard designs can be obtained from it. To prove this requires a return to the origin of the designs, in the realms of matrix theory, for there is a close tie up between Hadamard configurations and a class of matrices called *Hadamard matrices*, all of whose entries are ± 1. Roughly speaking, given a Hadamard matrix, change all − 1s to 0, and the resulting matrix is close to being the incidence matrix of a design. But that is an oversimplification.

Hadamard introduced his matrices when studying how large the determinant of a square matrix can be. Some restrictions are necessary before the problem makes sense, since clearly matrices of arbitrarily large determinant exist. So suppose the matrix is $n \times n$, and that each entry is numerically no larger than 1, i.e. $|a_{ij}| \leqslant 1$ for each choice of i, j. Then Hadamard proved that the determinant is at most $n^{\frac{1}{2}n}$, and that this value is actually obtained in magnitude if and only if each a_{ij} is ± 1 and $AA' = nI$. This result automatically leads to a study of such matrices.

DEFINITION 6.2. An $n \times n$ matrix is a *Hadamard matrix* of order n if $a_{ij} = \pm 1$ for each i, j, and $AA' = nI$.

Example 6.3.

$$\begin{bmatrix} 1 & 1 \\ -1 & 1 \end{bmatrix} \quad \text{and} \quad \begin{bmatrix} 1 & 1 & -1 & 1 \\ 1 & -1 & -1 & -1 \\ -1 & -1 & -1 & 1 \\ 1 & -1 & 1 & 1 \end{bmatrix}$$

are both Hadamard matrices.

Given a Hadamard matrix, it is permissible to interchange any two rows or any two columns, or to multiply any row or column by −1, for these operations do not effect the properties required by the definition. It is therefore possible to assume that a Hadamard matrix has its first row and its first column consisting entirely of + 1s. The matrix is then said to be *normalized*. The second matrix above can be normalized by multiplying the third row and the third column by −1.

The (i, j)th element of AA' is obtained by multiplying term by term the ith row of A with the jth column of A', i.e. the ith row of A with the jth row of A. If $i \neq j$, the value must be 0, and in this case the rows are said to be *orthogonal*. In the second example above, the rows are orthogonal since, for example, the first and second rows give

$$1.1 + 1.(-1) + (-1).(-1) + 1.(-1) = 0.$$

Now a sum of terms each of which is ± 1 can only be 0 if there is an even number of terms, so n must be even. But, further, if $n > 2$, n must be divisible by 4. For suppose that A is a normalized Hadamard matrix with $n > 2$. The first row has each entry + 1, and every other row, by orthogonality, must have $\frac{1}{2}n$ +1s and $\frac{1}{2}n$ −1s. The columns can be interchanged so that the second row consists of $\frac{1}{2}n$ +1s first, followed by $\frac{1}{2}n$ −1s. Now consider the third row (if $n > 2$). Suppose that the first $\frac{1}{2}n$ entries in it consist of u +1s and $(\frac{1}{2}n - u)$ −1s, and that the last $\frac{1}{2}n$ entries consist of v +1s and $(\frac{1}{2}n - v)$ −1s. Since the first and the third rows are orthogonal, it is necessary that the third row has $\frac{1}{2}n$ +1s, so

$$u + v = \tfrac{1}{2}n. \tag{6.12}$$

Since the second and third rows are also orthogonal,

$$u - (\tfrac{1}{2}n - u) - v + (\tfrac{1}{2}n - v) = 0,$$

i.e.

$$u - v = 0. \tag{6.13}$$

(6.12) and (6.13) give $u = v = n/4$, so that n must be a multiple of 4. The following result now is proved.

THEOREM 6.5. *If A is an n × n Hadamard matrix with n > 2, then n = 4m for some positive integer m. Further, each row has exactly 2m +1s and 2m −1s, and, for any two chosen rows, there are exactly m columns in which both rows have +1.*

The next step is to show how a new Hadamard matrix can always be obtained from a known Hadamard matrix. The proof is very simple. If H is $n \times n$ Hadamard, define a $2n \times 2n$ matrix K by

$$K = \begin{bmatrix} H & H \\ H & -H \end{bmatrix}.$$

For example, if H is the first matrix in Example 6.3, then

$$K = \begin{bmatrix} 1 & 1 & 1 & 1 \\ -1 & 1 & -1 & 1 \\ 1 & 1 & -1 & -1 \\ -1 & 1 & 1 & -1 \end{bmatrix}.$$

Then K is also Hadamard. Clearly all entries are ± 1, so all that has to be proved is that $KK' = 2nI$. Now each row of H consists of $\frac{1}{2}n$ +1s and $\frac{1}{2}n$ −1s, so each row of K contains n of each. Multiplying out KK', the diagonal elements are certainly all $2n$. Finally, use the orthogonality of the rows of H to deduce the orthogonality of the rows of K.

The implication of this result is that there is an unlimited supply of Hadamard matrices. These in turn yield an unlimited supply of symmetric block designs.

THEOREM 6.6. *Each normalized Hadamard matrix A of order $4m \geqslant 8$ yields a $(4m - 1, 2m - 1, m - 1)$-configuration.*

Proof. Delete the first row and the first column of A, and in what is left of A replace every -1 by 0. Denote the resulting matrix by B. B is a square matrix of order $(4m - 1)$, and each row has $(2m - 1)$ 1s and $2m$ 0s. Each row therefore represents a set of $(2m - 1)$ elements. Also, by Theorem 6.5, any two rows have 1s together in exactly $(m - 1)$ columns (not m, since the first column of A has been removed). Thus

$$BB' = mI + (m - 1)J$$

and this is sufficient to prove the theorem.

An example of this is included in the exercises below.

Exercises 6.3

1. Prove that, if $AA' = nI$, where A is $n \times n$, then $|\det A| = n^{\frac{1}{2}n}$.

2. If

$$B = \begin{bmatrix} 1 & 1 & 1 & 1 \\ 1 & -1 & 1 & -1 \\ 1 & 1 & -1 & -1 \\ 1 & -1 & -1 & 1 \end{bmatrix}$$

derive from B an 8 x 8 Hadamard matrix, and hence write down the incidence matrix of a $(7, 3, 1)$-configuration.

3. Fill in all the steps in the proof in the text that K is a Hadamard matrix.

4. Does condition (6.2) give any further information about m if a $4m \times 4m$ Hadamard configuration exists?

5. Using the matrix B of question 2 above, construct a 16 x 16 Hadamard matrix and deduce a $(15, 7, 3)$-configuration.

6. (a) If A is a $(0, 1)$-matrix and if B is the matrix obtained from A by replacing every 0 by -1, verify that $B = 2A - J$.
   (b) If A is the matrix of a (v, k, λ)-configuration, show, by evaluating BB', that B is a Hadamard matrix if and only if $v = 4(k - \lambda)$.

6.4. Error-correcting codes

The communication of information over a distance is subject to interference which sometimes leads to errors creeping in. As such interference is inevitable, steps must be taken to produce a system which will show up any errors present and, if possible, correct them.

The information is communicated by means of a code. This section will look at *binary codes*, in which each word is represented by a string of 0s and 1s of a fixed length, each such string representing a word in ordinary language. One way of looking at this is to think of the 0s as dots and the 1s as dashes, so that strings of 0s and 1s represent mathematically strings of short and long sound pulses. The general situation to be considered is the following, where errors arise during the transmission of the code words.

Message put into code
↓
Message transmitted
↓   ← possible error
Message received
↓
Message decoded.

Take a very simple situation to begin with. Suppose there are 16 code words, namely strings of 0s and 1s of length 4. The following message may be transmitted:

$$1\,1\,0\,0\,1\,1\,1\,0\,0\,0\,1\,1.$$

What will be received? Whether or not errors creep in, a string of 12 digits will be received, making three words, each of which is a code word. There will be no guarantee that they are the words which were transmitted, for there is no way of checking whether or not an error has crept in.

But suppose that each word is transmitted twice. The above message would be transmitted as

$$1\,1\,0\,0\,1\,1\,0\,0\,1\,1\,1\,0\,1\,1\,1\,0\,0\,0\,1\,1\,0\,0\,1\,1.$$

Here there is some check on the received message, for if the first four digits received differ from the next four, then an error has definitely arisen. For example, suppose that

$$1\,1\,0\,0\,0\,1\,0\,0\,1\,1\,1\,0\,1\,1\,1\,0\,0\,0\,1\,0\,0\,0\,1\,0$$

is received. It is known that an error has crept into the first word since 1 1 0 0 differs from 0 1 0 0. The error has been detected, but can it be corrected? There is no way of saying which is the correct one. Indeed, there is no saying definitely that either is the correct one! To see this, consider the last word transmitted. It was 0 0 1 1, transmitted twice, but 0 0 1 0 was received twice. The error is not detected. If the probability of an error occurring in any digit is small, it is unlikely that this identical repetition of the same error will occur, but it cannot be ruled out. All that the receiver can conclude is that the second word is most likely 1 1 1 0 and the third is most likely 0 0 1 0. As for the first, it is most likely one of 1 1 0 0 and 0 1 0 0, but there is no saying which. If, however, each word is transmitted thrice, not twice, and just one error occurs, the original word can be guessed using a majority decision principle. If the first word is received as

$$0 1 0 0 1 1 0 0 1 1 0 0$$

the error is spotted immediately, and the original is guessed to have been 1 1 0 0 on democratic (and probability) principles. It is the most likely candidate. The reader should now see why it is better to transmit each word an odd, rather than an even, number of times.

Some of the problems above arise from the fact that if even just one error occurs in transmission, the received word is another possible word in the code. But if the words of the code were chosen so that no two are close to each other, then this would be avoided. For example, if there were only four code words

$$0 0 0 0 0 0, \quad 0 1 0 1 0 1, \quad 1 0 1 0 1 0, 1 1 1 1 1 1$$

and if the first was transmitted but received as 0 0 0 0 0 1, then quite definitely an error has crept in since 0 0 0 0 0 1 could not have been transmitted. What was transmitted? Most likely it was 0 0 0 0 0 0 since this would involve just one error, whereas 0 1 0 1 0 1, for example, would involve two errors. Once this idea of making each code word sufficiently different from each other is grasped, it is no great leap to the idea of using block designs to construct codes. For with a (v, k, λ)-configuration, each row of the incidence matrix contains k 1s and $(v - k)$ 0s, and any two rows have 1s in the same λ columns. This means

$$
\begin{array}{cccc}
1\,1\,1 & 1\,1\,1 & 0\,0\,0 & 0\,0\,0 \\
1\,1\,1 & 0\,0\,0 & 1\,1\,1 & 0\,0\,0 \\
\underbrace{} & \underbrace{} & \underbrace{} & \underbrace{} \\
\lambda & k - \lambda & k - \lambda & v - 2(k - \lambda) - \lambda
\end{array}
$$

FIG. 6.2

that the rows differ in $2(k - \lambda)$ places (see Fig. 6.2), and therefore, if the rows are used as the code words, two words will not get confused unless a number of errors occur.

How different must the code words be before it is possible to correct a given number of errors?

THEOREM 6.7. *A code will detect all sets of h or fewer errors if any two words differ in at least* $(h + 1)$ *places.*

Proof. Even if h errors are made, the received string cannot be any of the code words.

THEOREM 6.8. *A code will correct all sets of h or fewer errors if any two words differ in at least* $(2h + 1)$ *places.*

Proof. What is required is that, if $g \leqslant h$ errors occur in transmission, the received string is nearer the original word than it is to any other word. If g errors are made, the resulting string will differ from any other word in at least

$$2h + 1 - g \geqslant 2h + 1 - h = h + 1 > h \geqslant g$$

places as required.

Example 6.4. Consider a $(4m - 1, 2m - 1, m - 1)$-configuration H. Take the rows of its incidence matrix A as the words of the code. They differ in at least $2(k - \lambda) = 2m$ places, and so the code will detect $(2m - 1)$ errors and correct $(m - 1)$ errors. The number of words available is only $(4m - 1)$. This number can be increased by noting that the complement $\bar{H}$ of the configuration (obtained from H by interchanging 0s and 1s in the incidence matrix A of H) is a $(4m - 1, 2m, m)$-configuration (see Exercises 6.1, question 4). The rows of its incidence matrix $\bar{A}$ also differ in $2m$ places. In how many places do the ith row of A and the jth row of $\bar{A}$ differ? They will differ in those places in which the ith row of A agrees with the jth row of A, and the number of such places is, by Fig. 6.2,

$$v - 2(k - \lambda) = 4m - 1 - 2m = 2m - 1.$$

Again, this is enough to correct $(m - 1)$ errors, so now $2(4m - 1)$ words are available. But, further, any row of A or $\bar{A}$ has $(2m - 1)$ 1s and $2m$ 0s, or vice versa, and so differs from the strings $0\,0\ldots0$ and $1\,1\ldots1$ in at least $(2m - 1)$ places. This gives two further code words, and now $8m$ words are available, detecting $(2m - 2)$ and correcting $(m - 1)$ errors.

This can be improved to $8m$ words detecting $(2m - 1)$ and correcting $(m - 1)$ errors by the following device. Precede each row of A by a 1, and each row of $\bar{A}$ by a 0. Then the ith row of A and the jth row of $\bar{A}$ differ in at least $2m$ places. Any two words in the resulting code therefore differ in at least $2m$ places, and the result now follows from the theorems.

Example 6.5. From the seven-point plane, a code which detects up to 3 errors and corrects 1 error can be constructed as in the previous example. The words are

```
1 1 1 1 1 1 1 1
1 1 1 0 1 0 0 0
1 0 1 1 0 1 0 0
1 0 0 1 1 0 1 0
1 0 0 0 1 1 0 1
1 1 0 0 0 1 1 0
1 0 1 0 0 0 1 1
1 1 0 1 0 0 0 1
0 0 0 1 0 1 1 1
0 1 0 0 1 0 1 1
0 1 1 0 0 1 0 1
0 1 1 1 0 0 1 0
0 0 1 1 1 0 0 1
0 1 0 1 1 1 0 0
0 0 1 0 1 1 1 0
0 0 0 0 0 0 0 0.
```

Any two words differ in at least 4 places, so that code will detect up to 3 errors and correct 1 error.

Exercises 6.4

1. The following sets form a $(11, 5, 2)$-configuration. Construct a code system of 24 words which will detect up to 5 errors and correct up to 2 errors.

$\{1, 3, 4, 5, 9\}$ $\{2, 4, 5, 6, 10\}$ $\{3, 5, 6, 7, 11\}$
$\{1, 4, 6, 7, 8\}$ $\{2, 5, 7, 8, 9\}$ $\{3, 6, 8, 9, 10\}$
$\{4, 7, 9, 10, 11\}$ $\{1, 5, 8, 10, 11\}$ $\{1, 2, 6, 9, 11\}$
$\{1, 2, 3, 7, 10\}$ $\{2, 3, 4, 8, 11\}$

2. Note that in the above design each set can be obtained from the previous one by adding 1 to each element (and replacing 12 by 1 if 12 results). Verify that the seven-point plane can also be obtained in this way starting with the set $\{1, 2, 4\}$. Such a design is said to be *cyclic*.

3. Using the idea of the previous question, construct a $(13, 4, 1)$-configuration from the set $\{1, 2, 4, 10\}$, and verify that it is a finite projective plane of order 3 (*the thirteen-point plane*). Use it to construct a code system of 26 words. How many errors will it detect? How many will it correct?

7 Steiner systems and sphere packings

7.1. Introductory remarks

IN recent years interest has been aroused in a family of combinatorial configurations called *Steiner systems* by a series of discoveries relating such systems to the theory of groups and the theory of sphere packings.

The basic problem in sphere packing is to find how to pack as many spheres as possible into a large volume, the spheres being assumed to be rigid, non-overlapping, and all of the same size. In three dimensions a sphere of radius r, centre the origin, is the set of points

$$\{(x_1, x_2, x_3): x_1^2 + x_2^2 + x_3^2 \leqslant r^2\},$$

and in general, a sphere of radius r in n dimensions, centre the origin, is defined to be

$$\{(x_1, x_2, \ldots, x_n): x_1^2 + \cdots + x_n^2 \leqslant r^2\},$$

so that, for example, a two-dimensional sphere is a circular disc. Although our three-dimensional existence denies us the experience of seeing spheres of four or more dimensions, the mathematics of such spheres is not much more difficult, and a number of combinatorial problems can be thought of intuitively as geometrical problems in three or more dimensions.

The problem of how best to pack circles in the plane has long been settled. Circles, all of the same size, are to be placed in the plane so that no two overlap, covering as large a proportion of the area as possible. Fig. 7.1

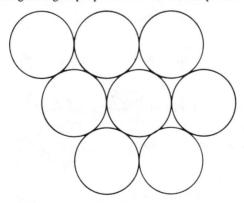

FIG. 7.1

shows the obvious way of fitting the circles together, and it is not too difficult to show that this is the best way of going about it (see [21]). By this packing, $\pi/\sqrt{12} = 0 \cdot 9069$ ... of the area of the plane is covered. The packing of Fig. 7.2, however, covers only $\pi/4 = 0 \cdot 785$ ... of the plane. Both of these packings have a regularity about them. In Fig. 7.1, the centres of the circles (assumed to be of radius $\frac{1}{2}$) are placed as in Fig. 7.3.

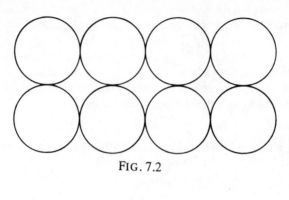

FIG. 7.2

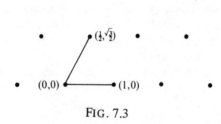

FIG. 7.3

If one of these points is chosen as the origin, the set Λ of centre points is given by

$$\Lambda = \left\{ m(1, 0) + n\left(\frac{1}{2}, \frac{\sqrt{3}}{2} \right) : m, n \text{ integers} \right\}.$$

Λ therefore has the following properties:
   (1) $\mathbf{0} \in \Lambda$.
   (2) $\mathbf{x} \in \Lambda \Rightarrow -\mathbf{x} \in \Lambda$.
   (3) $\mathbf{x} \in \Lambda$ and $\mathbf{y} \in \Lambda \Rightarrow \mathbf{x} + \mathbf{y} \in \Lambda$.
Such a set is called a lattice.

DEFINITION 7.1. A set of points x is a lattice if (1), (2), and (3) are satisfied.

Similarly, the packing in Fig. 7.2 arises from the lattice shown in Fig 7.4.

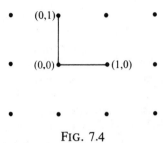

FIG. 7.4

$$\Lambda_0 = \{m(1, 0) + n(0, 1): m, n \text{ integers}\}$$
$$= \{(m, n): m, n \text{ integers}\}$$
$$= \text{set of points with integer coordinates.}$$

Λ_0, obviously a basic one, is called the *basic lattice* in two dimensions. Another example of a lattice in the plane is the set of all points (x, y) with x, y integers, either both even or both odd. The shortest distance between any two points of this lattice is $\sqrt{2}$, so a circle of radius $\frac{1}{2}\sqrt{2}$ can be centred at each point of the lattice to obtain a packing (Fig. 7.5).

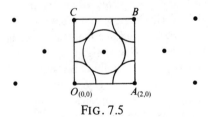

FIG. 7.5

The proportion of the plane covered can be found by considering $OABC$ which has area 4 units. The number of circles inside is two, one plus four quarters, covering an area of $2 . \pi . \frac{1}{2} = \pi$, so that a proportion of $\pi/4$ of the plane will be covered. This is the same as for the packing derived from Λ_0, as is to be expected since the packing of Fig. 7.5 is essentially that of Λ_0 turned through 45°. However, the idea of considering lattices with points whose coordinates are either all even of all odd is one to remember. Leech's lattice is such a one.

In three dimensions, experience has shown that there is no better packing than the obvious one. Take a large box and cover the bottom of it with a layer of spheres as in Fig. 7.1. Above this layer, a second layer is now to be formed by placing spheres into the hollows of the first layer. There is not, however, enough room to fit a sphere into every hollow, so fit one into every second hollow. These hollows are all the same distance apart as the centres of the spheres of the first layer were. Then add a third layer, the relationship of the third to the second being the same as that of the second to the first, and proceed in this way. Nature, usually so economical, makes use of this packing, but so far no-one has been able to prove that there is no better one. It fills $\pi/\sqrt{18} = 0\cdot7404\ldots$ of space, whereas the best result so far obtained is that it is impossible to fill more than $0\cdot7797\ldots$ of space (see Rogers [5]). The centres of the spheres in this present packing form the lattice of points

$$m(1, 0, 0) + n\left(\frac{1}{2}, \frac{\sqrt{3}}{2}, 0\right) + p\left(\frac{1}{2}, \frac{1}{2\sqrt{3}}, \frac{\sqrt{2}}{\sqrt{3}}\right),$$

with m, n, p integers.

A lattice consists of an array of points which continues in a regular pattern in all directions, the view from any one point being indistinguisable from the view from any other point. It should be remarked, however, that it is quite possible to have a certain regularity without having a lattice. If, for example, in the three-dimensional packing described above, the third layer was placed instead in the other hollows, so that the first and third layers are symmetric with respect to the second, and then further layers were added in either way, then a non-lattice packing would be obtained which also fills $\pi/\sqrt{18}$ of space. Both lattice and non-lattice packings occur naturally in certain crystals.

Packers of spheres in more than three dimensions have tended to concentrate on lattice packings for the obvious reason that the lattice pattern always helps to make things easier. Recently a lattice in twenty-four dimensions has been the subject of considerable interest due to its giving a very good packing result for spheres in twenty-four dimensions. This lattice, *Leech's lattice*, can be derived from a Steiner system, a configuration to be described in the pages ahead.

What makes one packing better than another? In two dimensions, the second and third packings to be considered lose space around each circle because circles fail to touch. In the first packing, each circle touches 6 others, whereas in the second and third each circle touches only 4 others. It is clearly reasonable to look for lattice packings in which each sphere touches as many others as possible. In packings in more than three dimensions, think not of the spheres but of the centres of the spheres, i.e. the points of the lattice. The distance between two lattice points $(x_1, x_2, \ldots, x_n)$ and $(y_1, y_2, \ldots, y_n)$ is $\{(x_1 - y_1)^2 + \cdots + (x_n - y_n)^2\}^{1/2}$, and if $b(\Lambda)$ denotes the shortest distance between any two points of the lattice Λ then a sphere of radius $\tfrac{1}{2}b(\Lambda)$ can be centred at each lattice point. The lattice packing will be a good one if there are many lattice points at distance $b(\Lambda)$ from each lattice point, since then many spheres will touch and valuable space will not be lost. This approach avoids complications about 'n-dimensional volume'.

Finally, a practical point. Since any point in a lattice is just as good as any other, always choose the sphere at the origin when calculating the number of spheres touching a given sphere. The required number will simply be the number of lattice points at distance $b(\Lambda)$ from the origin.

Exercises 7.1

1. Verify that the lattice packing of Fig. 7.1 covers $\pi/\sqrt{12}$ of the plane.

2. How many spheres touch any given sphere in the best known three-dimensional packings described in the text?

3. In n dimensions, let Λ denote the set of points whose coordinates are integers, either all even or all odd. Prove that Λ is a lattice, and that $b(\Lambda) = \sqrt{n}$.
   If now $n = 3$, and spheres of radius $\tfrac{1}{2}\sqrt{3}$ are centred at each lattice point, how many spheres will touch any given sphere?

4. In four dimensions, let Λ be as in the previous question. Show that there are 24 points of Λ at distance $b(\Lambda)$ from the origin. (Note that, unlike the situation in question 3, such points need not have all coordinates ± 1: for example, $(2, 0, 0, 0)$ is in Λ.) Thus Λ gives a packing in which each sphere touches 24 others. This is the best known in four dimensions.

5. In eight dimensions, let K be the set of all points whose coordinates are either given by one of the 16 rows exhibited in Example 6.5 or can be obtained from one of these rows by adding even numbers to each component. Show that K is a lattice, and find the smallest distance of a point of K from the origin. Show that there are 240 points of K at this distance from the origin. Once again, this is the best known result (see the table on p. 114).

7.2. Steiner systems

Recall that a (b, v, r, k, λ)-configuration consists of b subsets of a set of v elements, each subset having k elements, such that each pair of elements occurs together in exactly λ of the subsets, and each element occurs in exactly r of them. For such a design to exist, it is necessary to have

$$bk = rv \quad \text{and} \quad \lambda(v - 1) = r(k - 1).$$

In particular, choose $k = 3$ and $\lambda = 1$ to obtain a collection of three-element subsets, or *triples*, such that any given two elements appear together in one and only one of the triples. Then $3b = rv$ and $v - 1 = 2r$, so that, on eliminating r,

$$b = \frac{v(v - 1)}{6}.$$

Any such set of triples is called a *Steiner triple system*, and here is an example of one, corresponding to $v = 9$:

$$\{1, 2, 3\} \quad \{4, 5, 6\} \quad \{7, 8, 9\} \quad \{1, 4, 7\} \quad \{1, 5, 8\} \quad \{1, 6, 9\}$$
$$\{2, 4, 8\} \quad \{2, 5, 9\} \quad \{2, 6, 7\} \quad \{3, 4, 9\} \quad \{3, 5, 7\} \quad \{3, 6, 8\}.$$

The important point about the triples is that any pair of elements lies in exactly one triple, i.e. any two-element subset lies in exactly one of the three-element subsets. The generalization of this is to consider m-element subsets of a given set such that any l-element subset lies in exactly one of the m-element subsets.

DEFINITION 7.2. A *Steiner system* $S(l, m, n)$ is a collection of m-element subsets of an n-element set B such that every l-element subset of B lies in exactly one of the m-element sets. The set B is called the *base set*.

Example 7.1. The seven-point plane is an example of $S(2, 3, 7)$.

Example 7.2. The triple system listed above is an $S(2, 3, 9)$.

Example 7.3. A partition of a set of ab elements into b sets of a elements is an $S(1, a, ab)$.

A Steiner triple system is an $S(2, 3, n)$ for some n. The number of triples in such a system has been seen to be

$$\frac{n(n - 1)}{6} = \binom{n}{2} \Big/ \binom{3}{2}.$$

The generalization of this is

THEOREM 7.1. *The number of m-element sets in an S(l, m, n) is*

$$\binom{n}{l} \Big/ \binom{m}{l}.$$

Proof. There are $\binom{n}{l}$ *l*-element subsets of *B*. Each occurs in exactly one

m-element set. But each *m*-element set contains $\binom{m}{l}$ *l*-element sets. Thus

$$\binom{n}{l} = (\text{number of } m\text{-element sets})\binom{m}{l}.$$

Application

This result puts some restrictions on *l, m, n* if a system $S(l, m, n)$ is to exist. For example, there cannot be a system $S(2, 4, 10)$ since

$\binom{10}{2} \Big/ \binom{4}{2}$ is not an integer. Here is the kernel of an idea worth pursuing.

Can some other expressions be found which must also be integers? The answer lies in the following observation.

Suppose that there is a system $S(l, m, n)$, concentrate on one of the elements of the base set *B*, and consider only those *m*-element sets which contain it. If the chosen element is removed from each, a collection of $(m - 1)$-element subsets of an $(n - 1)$-element set is obtained, each $(l - 1)$-element subset lying in exactly one of the $(m - 1)$-element sets. Hence a system $S(l - 1, m - 1, n - 1)$ is obtained.

THEOREM 7.2. *If a system S(l, m, n) exists, so does a system S(l − 1, m − 1, n − 1). Consequently, so does a system S(l − 2, m − 2, n − 2) and in general a system S(l − u, m − u, n − u) for each u < l.*

This result gives yet more necessary conditions on *l, m, n* since not

only must $\binom{n}{l} \Big/ \binom{m}{l}$ be an integer, but so must $\binom{n-1}{l-1} \Big/ \binom{m-1}{l-1}$,

$\binom{n-2}{l-2} \Big/ \binom{m-2}{l-2}$ and so on. In particular, if a triple system $S(2, 3, n)$

exists, then $\binom{n}{2} \Big/ \binom{3}{2}$ and $\binom{n-1}{1} \Big/ \binom{2}{1}$, i.e. $n(n-1)/6$ and $(n-1)/2$

are both integers. The second of these requires that $n = 2u + 1$ for some integer u, and the first then requires that $(2u + 1)u/3$ must be an integer. This requires $u = 3s$ or $3s + 1$, whence $n = 6s + 1$ or $6s + 3$. As was shown by Kirkman in 1847, these conditions are in fact sufficient for a system $S(2, 3, n)$ to exist, and the reader can readily obtain a construction of a triple system $S(2, 3, n)$ for every n of the form $n = 6s + 1$ or $6s + 3$ by consulting Marshall Hall's book [2].

The most interesting Steiner systems in many ways are those with l as large as possible. But only four systems are known with $l > 3$! They are $S(5, 6, 12)$, $S(4, 5, 11)$, $S(5, 8, 24)$, and $S(4, 7, 23)$, the existence of those with $l = 5$ immediately implying the existence of the other two by Theorem 7.2. In the next section $S(5, 8, 24)$ will be singled out for special attention. The number of eight-element sets which form it is

$$\binom{24}{5} \bigg/ \binom{8}{5} = 759,$$ so at no stage will a list of these eight-element sets be

made. On the assumption that a system $S(5, 8, 24)$ exists (as it does—and it is unique), a number of its properties will be derived and then used. The reader who wishes to see a list of the 759 sets can satisfy his curiosity by consulting Todd's paper [22].

Exercises 7.2

1. Prove that a Steiner system $S(5, 7, 13)$ cannot exist.

2. How many five-element sets constitute $S(4, 5, 11)$? Using Theorem 7.2, state a value of n for which the existence of an $S(2, 3, n)$ can be deduced.

3. Show that the number of four-element sets in an $S(3, 4, 8)$ must be 14.

4. Construct a system $S(3, 4, 8)$ as follows. For seven of the four-element sets take the complements of the lines of the seven-point plane. For the other seven, add a new symbol '8' to the lines of the plane.

5. Verify that Theorem 7.2 does not exclude the possibility of the existence of a system $S(2, 7, 43)$. Prove, however, that such a system cannot exist. Theorem 6.4 may help.

6. Show that if a system $S(3, 4, n)$ exists, then n must be of the form $n = 6s + 2$ or $6s + 4$. (These conditions are known to be sufficient— see Hanani [14].)

7. Prove that any two of the four-element sets of a system $S(3, 4, n)$ can intersect in at most two elements. Deduce that if A is the incidence matrix of the system then any two rows of A will differ in at least four columns. The rows of A can therefore be used as a code which will detect up to 3 errors and correct 1 error.

7.3. $S(5, 8, 24)$

The idea of trying to obtain properties of a system $S(5, 8, 24)$ without knowing any of the 759 sets which appear in it may seem rather odd, but that is precisely what is going to be attempted. All that is given is that there is a twenty-four element set B and a collection of eight-element subsets of B, called *octads* such that each five-element subset of B lies in exactly one of the octads. The name 'octad' will be reserved for those eight-element subsets of B which are in the system: there are therefore many eight-element subsets of B which are not to be called octads. Of particular importance in subsequent investigations will be a knowledge of how many elements are in the intersection of two octads. But the starting point is the following result.

THEOREM 7.3. *In* $S(5, 8, 23)$,

(1) *the number of octads is 759,*

(2) *each element of B lies in 253 octads,*

(3) *each pair of elements lies in 77 octads,*

(4) *each triple of elements lies in 21 octads,*

(5) *each tetrad of elements (i.e. each four-element subset of B) lies in 5 octads,*

(6) *each quintuple of elements lies in just 1 octad.*

Proof. (1) has already been proved.

(2) The number of octads containing a given element is, by the proof of Theorem 7.2, the number of seven-element sets in $S(4, 7, 23)$, i.e.

$$\binom{23}{4} \bigg/ \binom{7}{4} = 253.$$

(3) Consider two elements x, y. The sets containing x form, when x is removed, a system $S(7, 4, 23)$. The sets of this system which contain y form, on the removal of y, a system $S(3, 6, 22)$. There are therefore

$$\binom{22}{3} \bigg/ \binom{6}{3} = 77 \text{ of them.}$$

(4) Similarly, $\binom{21}{2} \bigg/ \binom{5}{2} = 21.$

(5) $\binom{20}{1} \bigg/ \binom{4}{1} = 5.$

(6) This follows from the definition of $S(5, 8, 24)$.

At this point it will be helpful to have a few remarks about the ideas in the subsequent argument. If A and B are two subsets of a set S, the *symmetric difference* $(A + B)$ is defined to be the set of all elements of S which are in A or B but not both. $(A + B)$ is shaded in the diagram in Fig. 7.6. Clearly it is always true that $A + B = B + A$. The following are also easily verified:

$$A + A = \phi, \text{ the empty set;}$$

$$A + \phi = A;$$

$$A + (B + C) = (A + B) + C.$$

In algebraic language, the set of subsets of a set S form a *commutative group* under the operation $+$, with the empty set as the identity element, and with each set being its own inverse. Since there are 2^n subsets of a

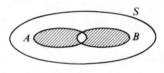

FIG. 7.6

set of n elements (Miscellaneous problems on Chapter 2), the subsets of the base set B of $S(5, 8, 24)$ form a group of 2^{24} members. Part of the work ahead will be to exhibit a subgroup of this group with 2^{12} members. Each octad will be in this group, but since there are only 759 octads, there will be other sets in it as well. The climax of the book will then be to show how this group quickly gives the famous Leech lattice which gives such good results in sphere packings. Most of the argument will be presented in full, but it will be necessary to leave a few gaps in the construction of the group.

If E and F are octads, $(E + F)$ will be in the group. How many elements does $(E + F)$ contain? This depends on the number of elements in $E \cap F$. For if $E \cap F$ has x members, then $(E + F)$ will have $2(8 - x)$ members (see Fig. 7.7). Thus there will be a chance of $(E + F)$ also being an octad if $x = 4$.

In the following proofs, Theorem 7.3 will be central.

PROPERTY 1. *No two octads can intersect in 5 or more elements.*

Proof. Otherwise Theorem 7.3, part (6), would be contradicted.

PROPERTY 2. *No two octads can intersect in exactly 3 elements.*

Proof. Let A be any octad, and let a_1, a_2, a_3 be any 3 of its elements. If it can be shown that there are 20 octads other than A which contain a_1, a_2, and a_3, and a fourth member of A, then by part (4) of Theorem 7.3, there will be no other octads left which also contain a_1, a_2, and a_3.

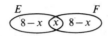

FIG. 7.7

Now there are precisely 5 four-element subsets of A which contain a_1, a_2, a_3. For the fourth element can be any one of the 5 remaining members of the octad A. Each of these four-element sets lies in 4 octads other than A. This gives rise to $5 \times 4 = 20$ octads, all containing a_1, a_2, a_3 and another element of A. These 20 octads are all different, for if the same octad arose in two different ways, say from the sets $\{a_1, a_2, a_3, a_4\}$ and $\{a_1, a_2, a_3, a_5\}$, then the quintuple $\{a_1, a_2, a_3, a_4, a_5\}$ would lie in that octad as well as in A, contradicting 7.3, part (6).

PROPRETY 3. *No two octads can intersect in exactly 1 element.*

Proof. Let A be any octad, and a_1 any element of A. It will be shown that no octad intersects A in just a_1. There are $\binom{7}{3} = 35$ four-element subsets of A containing a_1, and each of them lies in 4 octads other than A. This therefore gives rise to $35 \times 4 = 140$ octads containing a_1 and three other elements of A, and, as in the previous proof, these 140 octads must all be different. But a_1 lies in 253 octads altogether. Apart from A there are therefore another 112 octads containing a_1, and it will be shown that each of them intersects A in exactly 2 elements.

Let a_2 be any other member of A. a_1 and a_2 appear together in 76 other octads, but as part of a tetrad in $\binom{6}{2} \times 4 = 60$ octads, all these octads again being different. By property 2, there are therefore 16 octads left for which a_1 and a_2 will have to constitute their whole intersection with A. Similarly there 16 octads for a_1, a_3, 16 for $a_1, a_4, \ldots$, 16 for a_1, a_8. This accounts for $7 \times 16 = 112$ octads as required.

From the above three properties it follows that any two octads E and F intersect in 0, 2, or 4 elements.

$$|E \cap F| = 0 \Rightarrow |E + F| = 16,$$
$$|E \cap F| = 2 \Rightarrow |E + F| = 12,$$
$$|E \cap F| = 4 \Rightarrow |E + F| = 8.$$

If $E \cap F = \phi$, $E + F$ cannot possibly be an octad since it has 16 elements, but it turns out that its complement $(E + F)'$ is. Obviously $(E + F)'$ has $24 - 16 = 8$ elements, but remember that not every eight-element subset of the base set is one of the octads of the Steiner system. So there is still something to prove!

PROPERTY 4. *If E, F are octads and* $E \cap F = \phi$, *then* $(E + F)'$ *is also an octad.*

Proof. Suppose that $(E + F)' = \{a_1, \ldots, a_8\}$ is *not* an octad, and let C be the unique octad containing $a_1, \ldots, a_5$. Since C intersects E and F in even numbers of elements it must also intersect $(E + F)$ and hence $(E + F)'$ in an even number of elements. Thus C can be supposed to contain a_6 as well. C cannot contain a_7, since it would then have to contain a_8 (to make the intersection with $(E + F)'$ even) and this would result in C being equal to $(E + F)'$ which is not an octad. Thus

$$C = \{a_1, a_2, a_3, a_4, a_5, a_6, \quad , \quad \},$$
$$a_7 \notin C, \quad a_8 \notin C.$$

Let D be the unique octad containing a_1, a_2, a_3, a_4, a_7. D cannot contain a_5 or a_6 (otherwise $C = D$), so $a_8 \in D$, to make the intersection with $(E + F)'$ even. Thus

$$D = \{a_1, a_2, a_3, a_4, a_7, a_8, \quad , \quad \},$$
$$a_5 \notin D, \quad a_6 \notin D.$$

Finally, let G be the unique octad containing a_1, a_2, a_3, a_5, a_7. G must contain one of a_4, a_6, a_8 to make its intersection with $(E + F)'$ even. If $a_4 \in G$, then G and C have 5 elements in common, so that $G = C$. This requires that $a_7 \in C$, contradiction. Similarly if $a_6 \in G$. If $a_8 \in G$ then D and G coincide so that $a_5 \in D$, contradiction.

A similar argument now gives the following result.

PROPERTY 5. *If E and F are octads and* $|E + F| = 8$, *then* $(E + F)$ *is also an octad.*

This result can be expressed in a different way. Call two tetrads *complementary* if their union is an octad, and note that two complementary tetrads T_1, T_2 are necessarily disjoint. Then Property 5 states:

if two tetrads are both complementary to the same tetrad, then they are themselves complementary.

For $E = T_1 + T_3$, $F = T_2 + T_3$, and the property states that $(T_1 + T_2)$ is also an octad. (See Fig. 7.8.)

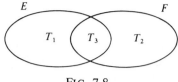

FIG. 7.8

The proof of Property 5 is left as an exercise for the reader. Properties 4 and 5 describe what $(E + F)$ is like when the intersection of E and F contains 0 or 4 elements: $(E + F)$ is an octad or the complement of an octad. There remains the problem of what happens when the intersection contains 2 elements, i.e. when $|E + F| = 12$. In this case, $D = E + F$ is called a *dodecad*. It turns out that these dedocads have the nice property that their complements, which of course have $24 - 12 = 12$ elements, are also dodecads. Further investigations show that there are 2576 dodecads.

Let K be the smallest subgroup of the group of subsets of the base set B under the operation of symmetric difference to contain all the octads. Since K is a group, it must contain all sets of the form $(E + F)$ where E and F are octads. It must also contain the complement of an octad D and hence must contain $D + D' = B$. It therefore contains the complement of every set H in K, since $H + B = H'$. K therefore contains all octads and their complements, and all dodecads and hence the following sets:

1	base set B
1	empty set
759	octads
759	complements of octads
2576	dodecads
4096 $= 2^{12}$.	

This remarkable appearance of the number 2^{12} is made all the more remarkable by the fact that these 2^{12} subsets of B constitute the whole of the group K. To show this it is necessary to prove that the symmetric difference of any two sets in the above list is also in the list. For example, if E, F are octads, then $(E + F)$ is in the list, as has already been shown. If E is an octad and F is the complement of an octad L, then

$$E + F = E + L' = E + \phi + L' = E + (L + L) + L'$$
$$= E + L + (L + L') = E + L + B = (E + L)',$$

where $(E + L)$, and hence its complement, is in the list. The cases involving dodecads are more difficult to prove, and the details are omitted. The reader who wishes further information and insight should study Todd's paper [22] which is, unfortunately, not easily accessible. This paper contains the material presented above, except that it does not explicitly mention the group structure. It was Conway [9] who pointed out this structure and used it to present Leech's lattice in a new way.

The next section will deal with the construction of the Leech lattice, assuming the group structure which has been explained in the past few pages. But before that, this section will be brought to an end with an explanation of how the number 2576 of dodecads is arrived at.

PROPERTY 6. *No dodecad contains an octad.*

Proof. Suppose that D is a dodecad containing an octad A. Then $D = E + F$ for some octads E, F, each of which has exactly 6 elements in common with D. Thus $A \neq E, A \neq F$. A cannot therefore have more than 4 elements in common with either E or F, but it has 8 elements altogether in common with E and F, so that it must have exactly 4 in common with each. Consequently, let $A \cap E = T_1, A \cap F = T_2$, where $T_1 \cap T_2 = \phi$. Now E and F have 2 elements in common, say a_1 and a_2.

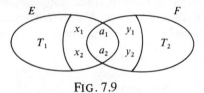

FIG. 7.9

Suppose further than E consists of T_1, a_1, a_2, x_1, x_2 and that F consists of T_2, a_1, a_2, y_1, y_2 (see Fig. 7.9). Then

$$T_1 \text{ and } T_2 \text{ are complementary} \tag{7.1}$$

and

$$T_1 \text{ and } \{a_1, a_2, x_1, x_2\} \text{ are complementary.} \tag{7.2}$$

By Property 5, second formulation, (7.1) and (7.2) imply that

$$T_2 \text{ and } \{a_1, a_2, x_1, x_2\} \text{ are complementary.} \tag{7.3}$$

But

$$T_2 \text{ and } \{a_1, a_2, y_1, y_2\} \text{ are complementary,} \tag{7.4}$$

and (7.3) and (7.4) now imply that

$$\{a_1, a_2, x_1, x_2\} \quad \text{and} \quad \{a_1, a_2, y_1, y_2\} \text{ are complementary.}$$

This last assertion is necessarily false since complementary tetrads are always disjoint. This contradiction means that the assumption that the dodecad D contains an octad A must have been false.

To count the dodecads, it is convenient to consider their structure (Fig. 7.10). Any dodecad D is the union of 2 six-element sets, or hexads, these hexads being *special* in the sense that they lie in an octad (not every six-element subset of the base set does). Any five-element subset of D lies in a unique

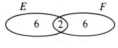

$$E \qquad\qquad F$$

FIG. 7.10

octad, and since this octad intersects D in an even number of elements, it must intersect D in exactly 6 elements. Thus the 12 elements of D have the property that any 5 of them determine a sixth, the 6 elements forming a special hexad. The number of special hexads contained in D is $\frac{1}{6}\binom{12}{5} = 132$, since each special hexad arises from 6 different quintuples. Now the total number of special hexads is $759 \times \binom{8}{6} = 759 \times 28$, so that, if it can be shown that each special hexad occurs in 16 dodecads, then it will follow that

$$\text{(number of dodecads)} . 132 = 759 . 28 . 16,$$

whence the number of dodecads is 2576.

Let R therefore be a special hexad. Since it lies in a unique octad, A, R can be extended to an octad by adding a further 2 elements in only one way. The number of dodecads containing R is the number of octads which intersect A in exactly these 2 elements. As in the proof of Property 3, this number is 16.

Exercises 7.3

1. Show that the symmetric difference of any two lines of the seven-point plane has 4 elements and, further, is the complement of one of the other lines.

2. Prove that $(A + B)' = A' + B = A + B'$ for any subsets A, B of a set S.

3. Provide a proof of Property 5. (Suppose that $(E + F)$ is not an octad and let C be the unique octad containing the first 5 members of $(E + F)$. Imitate the proof of Property 4.)

4. Construct a system $S(5, 6, 12)$ by taking as the base set B any dodecad of $S(5, 8, 24)$ and taking as the six-element sets the special hexads contained in B.

5. Consider the system $S(3, 4, 8)$ of Exercises 7.2, question 4, and say that 2 two-element sets are *complementary* if their union is one of the 14 four-element sets of the system. Show that if A, B are complementary and A, C are complementary, then B, C are also complementary (e.g. $A = \{1, 2\}, B = \{4, 8\}, C = \{3, 6\}$).

7.4. Leech's lattice

Let K be the group described in the previous section, consisting of all the octads of $S(5, 8, 24)$, their complements, the dodecads, the empty set and the base set $B = \{1, \ldots, 24\}$, under the operation of symmetric difference $+$. Then K has 2^{12} elements. If C is any member of K and m is any integer, let $C(m)$ denote the set of all twenty-four-dimensional vectors

$$\mathbf{x} = (x_1, x_2, \ldots, x_{24})$$

such that
(a) each x_i is an integer, and $\sum_i x_i = 4m$,
(b) $x_i \equiv m \pmod 4$ if $i \notin C$,
(c) $x_i \equiv m + 2 \pmod 4$ if $i \in C$.

Next define Λ to be the union of all the sets $C(m)$. Then Λ is Leech's lattice. To check that it is a lattice, it is necessary to verify that
(1) $\mathbf{0} \in \Lambda$,
(2) if $\mathbf{x} \in \Lambda$, then $-\mathbf{x} \in \Lambda$,
(3) if $\mathbf{x} \in \Lambda$ and $y \in \Lambda$, then $\mathbf{x} + \mathbf{y} \in \Lambda$.

Proof of (1). $\mathbf{0} = (0, \ldots, 0) \in \phi(0)$.
Proof of (2). If $\mathbf{x} \in C(m)$ then $-\mathbf{x} \in C(-m)$.
Proof of (3). If $\mathbf{x}$ and $\mathbf{y}$ both belong to Λ, then $x \in C(m)$ and $\mathbf{y} \in D(n)$ for some C, D, m, n. It will be shown that $\mathbf{x} + \mathbf{y} \in (C + D)(m + n)$ by considering the ith component, and it will follow that $\mathbf{x} + \mathbf{y} \in \Lambda$ since $C + D \in K$.

There are four cases to consider. First suppose that $i \in C$ and $i \in D$. Then $x_i \equiv m + 2$ and $y_i \equiv n + 2$, whence $x_i + y_i \equiv m + n$. But $i \notin C + D$, so $x_i + y_i \equiv m + n$ is what is required. The proofs of the other three cases are similar.

Λ is therefore a lattice, and a lattice packing of spheres can be obtained by centring at each point of Λ a twenty-four-dimensional sphere of radius $\frac{1}{2}b(\Lambda)$, where $b(\Lambda)$ is the shortest distance of a point of Λ from the origin, i.e. $b(\Lambda)$ is the minimum value of

$$(x_1^2 + \cdots x_{24}^2)^{1/2}$$

subject to the conditions

$$(x_1, \ldots, x_{24}) \neq (0, \ldots, 0)$$

and, for some integer m and set C,

$$\sum_i x_i = 4m,$$

$$x_i \equiv m \pmod 4 \quad \text{if } i \notin C,$$

$$x_i \equiv m + 2 \pmod 4 \quad \text{if } i \in C.$$

Note first of all that the last two conditions imply that either all the x_i are even or all the x_i are odd, according as m is even or odd. If C is an octad, sixteen of the x_i will be $\equiv m \pmod 4$ and eight will be $\equiv m + 2$. If C is the complement of an octad, then eight of the x_i will be $\equiv m$, and sixteen will be $\equiv m + 2$. If C is a dodecad there will be twelve of each, if $C = \phi$ all x_i will be $\equiv m$, and if $C = B$, all x_i will be $\equiv m + 2$. It follows from these statements that if, for example, all the x_i are even, and at least one x_i is not $\equiv 0 \pmod 4$, then at least eight of the x_i are non-zero. Eight of the x_i are then at least 2 in magnitude, so that

$$\sum_i x_i^2 \geqslant 8 . 2^2 = 32.$$

If, on the other hand, all the x_i are multiples of 4, m is even and so the condition $\sum_i x_i = 4m$ requires that either an x_i is at least 8, or there are at least two x_is which are at least 4 in magnitude. Hence again

$$\sum_i x_i^2 \geqslant 32.$$

The conclusion is that there is no point of Λ with coordinates all even which is at a shorter distance than $\sqrt{32} = 4\sqrt{2}$ from the origin. Also, there do exist points of Λ at this distance from $\mathbf{0}$: for example, take any octad C and choose $x_i = 2$ if $i \in C$, $x_i = 0$ otherwise.

Can a point of Λ with coordinates all odd be found closer to the origin? If all coordinates are odd and hence numerically $\geqslant 1$, then $\sum_i x_i^2 \geqslant 24$, whereas if even just one coordinate is numerically bigger than 1,

$$\sum_i x_i^2 \geqslant 3^2 + 23 . 1 = 32.$$

The only hope is therefore to find a point with every coordinate ± 1, and this turns out to be impossible. For the number of + 1s would have to be 0, 8, 12, 16, or 24, and the corresponding sums $\sum_i x_i$ would then be $-24, -8, 0, 8$, or 24, all of which are of the form $4m$ with m even. But m must be odd to make each x_i odd.

It follows that $b(\Lambda) = \sqrt{32} = 4\sqrt{2}$, and that the corresponding packing is achieved by centring a sphere of radius $2\sqrt{2}$ at each point of Λ. To show that this is a particularly good packing, it will be shown that each sphere touches a very large number of other spheres. In finding the best possible packing in any given number of dimensions, a two-way sandwiching procedure is carried out. In one direction an attempt is made to construct a packing with each sphere touching as many others as possible. In the other direction, theoretical arguments are used to prove that no sphere could possibly touch more than a certain number of spheres. The closer these two numbers are together, the better the constructed lattice packing is considered to be. Let $a(n)$ be the largest number of spheres touching a given sphere in any packing yet constructed in n dimensions. We shall compare $a(n)$ with the upper bound $d(n)$ proposed by Coxeter [10] and evaluated numerically by Leech. This bound has not yet been proved, but is considered to be certainly true. It will act as an excellent estimator of how good $a(n)$ is. The following results are known (see [18]).

n	$a(n)$	$d(n)$	$\dfrac{a(n)}{d(n)}$
2	6	6	1·00
3	12	12	1·00
4	24	26	0·92
⋮			
8	240	244	0·98
9	306	401	0·76
⋮			
12	840	1 637	0·51
⋮			
20	17 400	51 421	0·34
⋮			
23	93 150	175 696	0·53
24	196 560	263 285	0·75

For $n \leqslant 8$, all ratios are at least 0·82, whereas the ratio which Leech's lattice gives for twenty-four dimensions is the best ratio for any $n > 9$; hence its importance.

It therefore remains to show that $a(24) = 196\ 560$. To do this it must be shown that there are 196 560 points $(x_1, \ldots, x_{24})$ of Λ for which $\sum_i x_i^2 = 32$. The cases of even and odd x_i will be considered separately.

Even case

One way of having $\sum_i x_i^2 = 32$ is to have eight of the x_i equal to ± 2, and all the other x_i zero. The eight ± 2s will be in places corresponding to an octad. There are therefore 759 ways of choosing where the 2s go, and it remains to consider how many arrangements of signs $+$, $-$ are possible for each such placing of the 2s. The condition $\sum_i x_i = 4m$ must be satisfied with m even, and this requires an even number of $+2$s and an even number of -2s. The first seven 2s can therefore be given any sign (do this in 2^7 ways), and the final sign will then be determined uniquely. Thus $759 \cdot 2^7 = 97\ 152$ points of arise in this way.

Another way would be to have some x_is numerically equal to 4. No x_i can then be 2, for there would then have to be at least eight 2s, contributing 32 to $\sum_i x_i^2$, leaving no room for a 4. There must therefore be exactly two 4s and 22 0s. All x_is are therefore divisible by 4, so the corresponding set C must be ϕ or B.

If $C = \phi$, then $m \equiv 0$ and so $\sum_i x_i = 0$, with one $+4$ and one -4. These can be placed in $24 \cdot 23$ ways. If $C = B$, then $m \equiv 2$ and so $\sum_i x_i = 8$ or -8. There are therefore either two $+4$s or two -4s. Two $+4$s can be placed in $\binom{24}{2}$ ways, as can two -4s. The total number of possibilities is therefore $2 \cdot \binom{24}{2} + 24 \cdot 23 = 1104$.

No other possibilities exist with even coordinates since $x_i = 6$ gives $x_i^2 = 36 > 32$.

So far $1104 + 97\ 152 = 98\ 256$ points of Λ have been exhibited.

Odd case

It has already been seen that one ± 3 and twenty-three ± 1s give $\sum_i x_i^2 = 32$, and that this is the only possibility. All that remains is to decide where the 3 can be placed and what signs can be chosen.

Suppose that one of the coordinates is +3, and that there are n -1s and $(23 - n)$ +1s. Then $\sum_i x_i = 26 - 2n = 4m$ for some odd integer m.

Further, by the conditions (b) and (c) in the definition of Leech's lattice, the number of coordinates which are $\equiv 1$ (mod 4) must be 0, 8 12, 16, or 24. 24 is clearly impossible since one coordinate is +3. If the number is 8, n is 15, so that $m = -1$. Thus $x_i \equiv m + 2$ for eight values of i, so the corresponding set C is an octad. Again, if the number is 16, then $n = 7$ and $m = 3$. Thus $x_i \equiv m + 2$ for sixteen values of i, so the corresponding set C is the complement of an octad. For each of the 759 octads there is a choice of sixteen places where the 3 might appear, similarly there are eight choices if C is the complement of an octad. This gives $24 . 759$ choices of the point $\mathbf{x}$ of Λ altogether, and this is doubled when those with -3 are considered. Thus $2 . 24 . 759$ points at distance $b(\Lambda)$ from $\mathbf{0}$ have so far been found.

If the number of coordinates which are $\equiv 1$ (mod 4) is twelve then $n = 11$ and $m = 1$. Then $x_i \equiv m + 2$ for twelve of the is, so that the corresponding set C is a dodecad. For each dodecad the 3 can be placed in twelve ways, and since the same analysis holds for vectors with a -3, the dodecads therefore give $24 .$ (number of dodecads) points of Λ.

If the number is zero, there must be a 3 and 23 -1s, giving $m = -5$. Here $x_i \equiv m + 2$ for no value of i, so the corresponding set C must be ϕ. With twenty-four choices of where to put the 3, twenty-four points of Λ are obtained. With -3, another twenty-four obtained.

Thus, altogether,

$24 .$ (2 + number of dodecads + number of octads and their complements)

$$= 24 . 2^{12} = 98\ 304$$

points of Λ with odd coordinates have been found. The total number of points of Λ at distance $\sqrt{32}$ from the origin is therefore $98\ 256 + 98\ 304 = 196\ 560$. This verifies the entry in the table.

But many problems still remain unsolved. Is this the best possible packing in twenty-four dimensions? Are there larger Steiner systems, and if so, can they be used to construct particularly good packings in higher dimensions? So little is known about packings that not even the case of three dimensions has been fully dealt with. No mathematician can be satisfied with this situation. Is there a simple idea still to be stumbled upon which will open up the unknown?

Such is the challenge, and the beauty, of combinatorial mathematics.

Solutions to exercises

Exercises 1.1.
2. $a_n = (n!)^2$ 3. $a_n = \dfrac{2}{n}$
4. (a) $\frac{1}{2}n(n+1)$,   (b) $\frac{1}{2}(n+2)(n-1)$

Exercises 1.2.
1. $220, 715, 2002$ 4. 462
5. $h(5, 10) = 126$

Exercises 2.1.
1. $9!, 5 \times 8!$ 2. $5! \times 22!$
3. $12! - 2.11! = 10.11!$ 5. $\frac{1}{2}(n-1)!$

Exercises 2.2.
1. $840, 56, 15\ 120$ 2. $26^3 \times 999$
3. $\dfrac{75!}{55!}$ (taking order into account)
4. (a) 2.5^3, (b) $2.4!$ 5. $12!$
6. $k(k-1)$ 7. (a) 26^5 (b) $\dfrac{26!}{21!}$
8. 2^n

Exercises 2.3.
1. $1 + 8x + 28x^2 + 56x^3 + 70x^4 + 56x^5 + 28x^6 + 8x^7 + x^8$,
   $1 - 8x + 28x^2 - 56x^3 + 70x^4 - 56x^5 + 28x^6 - 8x^7 + x^8$
2. $330, 1716, 6435$
3. $1 + 8x + 36x^2 + 120x^3 + \cdots$
4. $f(3, 10) = 66$ 5. 36 (see Exercises 1.2, question 5)
6. (a) $\dbinom{10}{4} \times \dbinom{15}{4}$,   (b) $\dbinom{10}{5}\dbinom{15}{3} + \dbinom{10}{6}\dbinom{15}{2} + \dbinom{10}{7}\dbinom{15}{1} + \dbinom{10}{8}$,
   (c) $\dbinom{25}{8} - \dbinom{15}{8} - 10.\dbinom{15}{7}$.
7. $\dbinom{14}{7}$ 11. $\dbinom{15}{11}, \dbinom{14}{10}$

Exercises 2.4.
1. $1 + \frac{1}{2}x + \frac{3}{8}x^2 + \frac{5}{16}x^3 + \cdots$
2. $(1+x)^{1/4} = 1 + \dfrac{1}{4}x - \dfrac{3}{4.4.2}x^2 + \dfrac{7}{4.4.4.2}x^3 - \cdots$
   $(1-x)^{1/4} = 1 - \dfrac{1}{4}x - \dfrac{3}{4.4.2}x^2 - \cdots$

$$(1 + x)^{-1/4} = 1 - \frac{1}{4}x + \frac{5}{4.4.2}x^2 - \frac{5.9}{4.4.4.6}x^3 + \cdots$$

$$(1 - x)^{-1/4} = 1 + \frac{1}{4}x + \frac{5}{4.4.2}x^2 + \cdots$$

Miscellaneous Problems on Chapter 2
1. 24
2. (1) $4^5 + 6^3 . 4^2$ (2) $4^2 6^3 + 4 . 6^3$
3. (a) $f(3, 8) = 45$, (b) $f(4, 18) = 1330$
4. (a) 12^7, (b) $\frac{12!}{5!}$
6. $5!, 5! - 2.4!$
10. 24

Exercises 3.1.
1. $\frac{10!}{2^5 . 5!} = 945$
2. (a) $\frac{16!}{8!}$, (b) $\frac{16!}{2^8 . 8!}$
3. $\frac{52!}{(13!)^4}$
4. $\frac{16!}{(4!)^5}$, $\binom{14}{2}\frac{12!}{(4!)^3 . 3!}$
5. $(1, 3), (2, 5), (4, 6), (7, 8)$

Exercises 3.2.
1. $1, 4, 3, 2$
2. a, b, d, e, c, g, f

Exercises 3.3.
1. aE, bC, cB, dD, eA.
4. $G_1B_3, G_2B_2, G_3B_4, G_4B_1$
5. $c_1L_2, c_2L_6, c_3L_4, c_4L_1, c_5L_5$

Exercises 3.4.
1. $\{1, 2, 3, 5\}$
2. $\{1, 3, 4, 5\}$

Exercises 4.1.
1. $u_5 = 9$
2. $a_4 = 9$

Exercises 4.2.
1. $a_n = (n - 1)2^n$
2. $a_n = 2^n - 3^{n-1}$
6. $a_n = 2a_{n-1} + 3a_{n-2}$
9. $D_n = n + 1$ if $a^2 = 1$
10. $a_n = (n + 1)$th Fibonacci number
11. $b_n = \frac{1}{2}\{(1 + \sqrt{2})^{n+1} + (1 - \sqrt{2})^{n+1}\}$

Exercises 4.3.
10. Sequence is $\{b_n\}$, where $b_1 = a_1$, $b_n = a_n - a_{n-1}$ $(n \geqslant 2)$.

12. $a_n = \left(\dfrac{n+2}{2}\right)^2$ if n even, $= \dfrac{(n+1)(n+3)}{4}$ if n odd.

Number of sequences is $\frac{1}{2}(3^n + 1)$.

Exercises 4.4.
2. $a(10, 5) = 5 - 10 \cdot 2^{10} + 10 \cdot 3^{10} - 5 \cdot 4^{10} + 5^{10}$
   $= 5\ 103\ 000$

Exercises 5.1.
1. Divide (4.22) by $n!$ and let $n \to \infty$
2. 47   3. 520   4. 4 per cent
6. $5!$, 44, 45 = 5 x number of derangements of 4 numbers.
7. $24\ 024$

Exercises 5.2.

1. $\displaystyle\sum_{k=0}^{8} \binom{8}{k} \dfrac{8!}{(8-k)!} x^k$

2. (a) $1 + 8x + 15x^2 + 9x^3$,   (b) $1 + 6x + 6x^2$
3. 132   4. 13   7, 30   9. 12

Exercises 6.1.
2. $A'A = AA' = 2I + J$
3. $r(k - 1) \neq \lambda(v - 1)$

Exercises 6.2.
1. The proof of Theorem 6.2 shows that $A'A$ has non-zero determinant. Alternatively, from $A'A = (k - \lambda)I + \lambda J$ and $JA = kJ$ it

   follows that $\left(A' - \dfrac{\lambda}{k} J\right) A = (k - \lambda)I$, whence

   $A^{-1} = \dfrac{1}{k - \lambda}\left(A' - \dfrac{\lambda}{k} J\right)$.

4. Incidence matrix is

$$
\begin{matrix}
1 & 0 & 0 & 0 & 1 & 0 & 1 \\
1 & 1 & 0 & 0 & 0 & 1 & 0 \\
0 & 1 & 1 & 0 & 0 & 0 & 1 \\
1 & 0 & 1 & 1 & 0 & 0 & 0 \\
0 & 1 & 0 & 1 & 1 & 0 & 0 \\
0 & 0 & 1 & 0 & 1 & 1 & 0 \\
0 & 0 & 0 & 1 & 0 & 1 & 1
\end{matrix}
$$

Exercises 6.3.

2. Incidence matrix is

$$\begin{matrix}
0 & 1 & 0 & 1 & 0 & 1 & 0 \\
1 & 0 & 0 & 1 & 1 & 0 & 0 \\
0 & 0 & 1 & 1 & 0 & 0 & 1 \\
1 & 1 & 1 & 0 & 0 & 0 & 0 \\
0 & 1 & 0 & 0 & 1 & 0 & 1 \\
1 & 0 & 0 & 0 & 0 & 1 & 1 \\
0 & 0 & 1 & 0 & 1 & 1 & 0
\end{matrix}$$

4. No. $\lambda(v - 1) = k(k - 1)$ becomes $(m - 1)(4m - 2) = (2m - 1)(2m - 2)$ which is true for every value of m.

Exercises 6.4.

1. Follows Example 6.5.
3. First few rows of incidence matrix A are

$$\begin{matrix}
1 & 1 & 0 & 1 & 0 & 0 & 0 & 0 & 0 & 1 & 0 & 0 & 0 \\
0 & 1 & 1 & 0 & 1 & 0 & 0 & 0 & 0 & 0 & 1 & 0 & 0 \\
0 & 0 & 1 & 1 & 0 & 1 & 0 & 0 & 0 & 0 & 0 & 1 & 0 \\
0 & 0 & 0 & 1 & 1 & 0 & 1 & 0 & 0 & 0 & 0 & 0 & 1 \\
1 & 0 & 0 & 0 & 1 & 1 & 0 & 1 & 0 & 0 & 0 & 0 & 0
\end{matrix}$$

For code, take the rows of this matrix, and their complements. It will detect up to 5 errors and correct up to 2. For

rows of A differ in at least 6 places,
rows of A differ in at least 6 places,
row of A and row of A differ in at least 7 places.

Exercises 7.1.

2. 12 (6 in same layer, 3 in the layers above and below)
3. 8   4. $b(K) = 2$

Exercises 7.2.

2. 66, $n = 9$
5. Any $S(2, 7, 43)$ would be a finite projective plane of order 6.

Bibliography

THE following books are mentioned in the text. Of these, Hall's is recommended for further reading in the subject, as is Ryser's.

1. FISHER, R. A. *The design of experiments.* Oliver and Boyd, Edinburgh (1935).
2. HALL, M. *Combinatorial theory.* Blaisdell, Waltham, Mass. (1967).
3. HARARY, F. *Graph theory.* Addison–Wesley, New York (1969).
4. MIRSKY, L. *Transversal theory.* Academic Press, New York (1971).
5. ROGERS, C. A. *Packing and covering.* Cambridge University Press (1964).
6. RYSER, H. J. *Combinatorial mathematics.* Wiley, New York (1963).
7. WILSON, R. J. *Introduction to graph theory.* Oliver and Boyd, Edinburgh (1972).

The following papers were also referred to.

8. ANDERSON, I. Perfect matchings of a graph, *J. comb. Theory* **10**, 183–6 (1971).
9. CONWAY, J. H. A group of order 8, 315, 553, 613, 086, 720,000, *Bull. London math. Soc.* **1**, 79–88 (1969).
10. COXETER, H. S. M. An upper bound for the number of equal non-overlapping spheres that can touch another of the same size, *Proc. Symposia pure math.* **7**, 53–71 (1963).
11. FISHER, R. A. An examination of the different possible solutions of a problem in incomplete blocks, *Ann. Eugen.* **10**, 52–7 (1940).
12. GALE, D. Optimal assignments in an ordered set: an application of matroid theory, *J. comb. Theory* **4**, 176–80 (1968).
13. HALL, P. On representatives of subsets, *J. London math. Soc.* **10**, 26–30 (1935).
14. HANANI, H. On quadruple systems, *Can. J. Math.* **12**, 145–57 (1960).
15. HARARY, F. and READ, R. C. The enumeration of tree-like polyhexes, *Proc. Edinburgh math. Soc.* **17**, 1–14 (1970).
16. KIRKMAN, REV. T. On a problem in combinations, *Cambr. Dublin Math. J.* **2**, 191–204 (1847).
17. LEECH, J. Notes on sphere packings, *Can. J. Math.* **19**, 251–67 (1967).
18. LEECH, J. and SLOANE, N. J. A., Sphere packings and error correcting codes, *Can. J. Math.* **23**, 718–45 (1971).
19. MIRSKY, L. and PERFECT, H. Systems of representatives, *J. Math. Analysis Applic.* **15**, 520–68 (1966).
20. RIORDAN, J. and SHANNON, C. E. The number of two-terminal series-parallel networks, *J. math. Phys.* **21**, 83–93 (1942).
21. SERGE, B. and MAHLER, K. On the densest packing of circles, *Amer. math. Mon.* **51**, 261–70 (1944).

22. TODD, J. A. A representation of the Mathieu group as a collineation group. *Annali Mat. pura appl.* **71**, 199–238 (1966).

23. TUTTE, W. T. The factorisation of linear graphs, *J. London math. Soc.* **22**, 107–111 (1947).

24. WOODALL, D. R. A market problem, *J. comb. Theory* **10**, 275–87. (1971).

Index

algorithm, 23, 25, 36
assignment problems, 25, 29, 34, 57, 71
base set, 102
benzene rings, 52
binary code, 92
binary sequence, 11, 45, 58, 92
binomial coefficient, 14
binomial theorem, 4, 14, 18, 46
block design, 77
Bruck–Ryser theorem, 85

Celtic, 24
chessboard, 66
circular arrangements, 9, 19, 76
complement of a design, 82
Conway, J. H. 110
Coxeter's bound, 114
cyclic design, 96

degree of a vertex, 23
derangement, 41, 55, 65
distinct representatives, 25
dodecad, 109
dual of a design, 88

error-correcting codes, 92
exchange property, 35
exponential function, 15, 49

Fibonacci sequence, 2, 38, 42, 51
finite projective plane, 78, 84
Fisher's inequality, 77, 80

Gale, D. 34
generating function, 5, 45, 62
golf-ball problem, 1, 4, 5, 47, 54
graph, 23
group, 106

Hadamard configuration, 88
Hadamard matrix, 89
Hall's theorem, 25, 33
hexagons, 52

incidence matrix, 78
inclusion–exclusion principle, 63
independent matrix elements, 30
interfering squares, 68

Kirkman, Rev. T. 104

Konig–Egerváry theorem, 32
latin square, 27, 74
lattice, 99
Leech's lattice, 99, 100, 112

marriage problem, 26
matroid, 35, 37
max–min theorem, 32
ménage, problème de, 76

network, 59
normalized matrix, 89

octad, 105
optimal assignments, 29, 34
orthogonal, 89

packing of spheres, 97
pairings, 21
partition, 50, 53, 61
Pascal's triangle, 13, 16
perfect matching, 23
permutation, 8, 58

rabbits, 42
Rangers, 24
recurrence relation, 2, 38, 54
Riordan, J. 59
rook polynomial, 67
Ryser, H. 85

self-dual, 88
seven-point plane, 78, 95
Shannon, C. E. 59
simple electrical network, 59
simple rooted tree, 38, 45
sphere in n dimensions, 97, 101
Steiner system, 101
Stirling numbers, 58
symmetric design, 80, 83
symmetric difference, 106
$S(5,8,24)$, 105

tetrad, 105
thirteen-point plane, 84, 96
tree, 38
triple system, 102
Tutte, W. T. 37

varieties, 77
vertex, 23

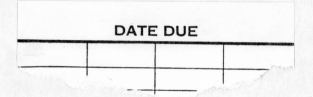

DATE DUE